Palgrave Studies in Literature, Science and Medicine

Series Editors
Sharon Ruston
Department of English and Creative Writing
Lancaster University
Lancaster, UK

Alice Jenkins
School of Critical Studies
University of Glasgow
Glasgow, UK

Catherine Belling
Feinberg School of Medicine
Northwestern University
Chicago, IL, USA

Palgrave Studies in Literature, Science and Medicine is an exciting new series that focuses on one of the most vibrant and interdisciplinary areas in literary studies: the intersection of literature, science and medicine. Comprised of academic monographs, essay collections, and Palgrave Pivot books, the series will emphasize a historical approach to its subjects, in conjunction with a range of other theoretical approaches. The series will cover all aspects of this rich and varied field and is open to new and emerging topics as well as established ones.

More information about this series at
http://www.springer.com/series/14613

Michelle Geric

Tennyson and Geology

Poetry and Poetics

Michelle Geric
University of Westminster
London, UK

Palgrave Studies in Literature, Science and Medicine
ISBN 978-3-319-66109-4   ISBN 978-3-319-66110-0   (eBook)
DOI 10.1007/978-3-319-66110-0

Library of Congress Control Number: 2017950393

Cover illustration: Simon Larbalestier/Alamy Stock Photo

Printed on acid-free paper

This Palgrave Macmillan imprint is published by Springer Nature
The registered company is Springer International Publishing AG
The registered company address is: Gewerbestrasse 11, 6330 Cham, Switzerland

For my father,
Alexander Lubo Geric (1925–2005)
"...he for whose applause I strove"
(In Memoriam, Tennyson)

ACKNOWLEDGEMENTS

I should like to thank Cambridge University Press for permission to use a version of my article published by them in *Victorian Literature and Culture* (14.42, 59–79) (Chap. 5). I also thank West Virginia University Press for their permission to use a version my article originally published in *Victorian Poetry* (51.1, 2013, 37–62) (Chap. 6). I thank members of staff at Birkbeck College Library, Senate House Library (University of London), the British Library, and particularly Grace Timmins at the Tennyson Research Centre in Lincoln. I am very grateful for the generosity of Michael A. Taylor and Lyall I. Anderson of Leicester University for allowing me to read their "Tennyson and the Geologists. Part 2. Saurians and the Isle of Wight" (*Tennyson Research Bulletin* 10.5, in press) before publication, and for Michael Taylor's generosity and openness. I thank all friends and colleagues in the Department of English, Linguistics and Cultural Studies at the University of Westminster for the great atmosphere they create, and I thank the department for its support in funding and organising the research leave that enabled me to work on this book. I should particularly like to thank Alexandra Warwick and Simon Avery at the University of Westminster for reading my work and for their comments on various chapters. I am indebted to Martin Willis, not only for his help and support for this project but more generally for his untiring support of scholarship in literature and science. I should also like to thank Sue Wragg for her friendship and encouragement, Mike Sanders, whose inspirational teaching was responsible for my interest in nineteenth-century studies, and Liz Cashden, a phenomenal teacher and poet. My very special thanks go to Deirdre McFeely and Emma McEvoy—I am very grateful to Emma as both colleague and friend for her comments on chapters and for her advice on many different issues, and to Deirdre, who has been a valued friend and reader of my work for many years—their friendship and intellectual support have been immensely important to me.

Finally, I thank my family; Deborah and Sam Torpey, Jeanne Heaslewood, Valerie Coles, my mother Katherine Gordon, my sister Leyla Geric—for all those childhood poetry readings, and so much more—and Richard Barker whose unfailing support, patience and advice has been invaluable.

Abbreviated References

After the first citing, references to the following works will be included parenthetically in the main body of the text by abbreviated title, volume and page number as appropriate.

Dialogic M. M. Bakhtin, *The Dialogic Imagination*, ed. Michael Holquist, trans. Caryl Emerson and Michael Holquist (Austin, Texas: University of Texas Press, 1981).

HIS William Whewell, *History of the Inductive Sciences* (1837) (London: Frank Cass, 1967), 3 vols.

Letters ALT *Letters of Alfred Lord Tennyson*, ed. Cecil Y. Lang and Edgar F. Shannon, Jr. (Oxford: Clarendon Press, 1982), 2 vols.

Memoir Hallam T. Tennyson, *Alfred Lord Tennyson: A Memoir by his Son* (London: Macmillan, 1897), 2 vols.

ORS Hugh Miller, *The Old Red Sandstone, or, New Walks in an Old Field* (Edinburgh: John Johnstone, 1841).

PG Charles Lyell, *Principles of Geology*, (1830–3) (Chicago: University of Chicago Press, 1990), 3 vols.

Problems Mikhail Bakhtin, *Problems of Dostoevsky's Poetics*, ed. and trans. Carl Emerson (Manchester: Manchester University Press, 1984).

Remains Arthur Hallam, *Remains in Verse and Prose of Arthur Henry Hallam* (London: W. Nicol, 1834).

Words Richard Chenevix Trench, *On the Study of Words* (1851) (London: John W. Parker and Son, 1853).

CONTENTS

CHAPTER 1

Introduction: Between a Rock and a Hard Place

no other object in the universe dominates human perception to the extent of the Earth. This dominion is all the more powerful because it is unperceived; the Earth provides the fabric on which all experience is located.
Robert Muir Wood, *The Dark Side of the Earth* (1985), 7.

Tennyson had an enduring interest in geology, which has long been known by students and scholars of literary studies. As part of a broad introduction to nineteenth-century literature, students routinely study *In Memoriam*'s 'geological stanzas' as evidence of both Tennyson's engagement with geology and its significance for Victorian religious belief; those "dreadful hammers" of which Ruskin complained.[1] The role played by geology in Tennyson's poetic imagination—taking *In Memoriam* as the usual example—has had much attention. Excavated dinosaur fossils recovered from "scarped cliff and quarried stone" reveal how 'Nature' "care[s] for nothing", while the "hills" that are "shadows" and that "flow / From form to form" demonstrate Tennyson's knowledge of contemporary ideas of geomorphology and specifically

---

[1] John Ruskin, *The Works of John Ruskin* (1912), Letter to Henry Acland (1851), 115.

© The Author(s) 2017
M. Geric, *Tennyson and Geology*, Palgrave Studies in Literature, Science and Medicine, DOI 10.1007/978-3-319-66110-0_1

1

the uniformitarianism expounded by Charles Lyell in his *Principles of Geology* (1830–1833). Tennyson's knowledge of Lyell's *Principles* and its significance for his poetic thinking has been documented by literary critics for at least the past sixty years.[2] However, Lyell's geology, and geology generally, has nearly always been read in conjunction with *In Memoriam* (1851), and in consequence, the extent to which it figures in other poems has not yet been fully appreciated. This study examines Tennyson's treatment of geology over an extended period across his three major mid-century poems, *The Princess* (1847), *In Memoriam* and *Maud* (1855). It argues that Tennyson's poetics specifically and consistently engaged with geological patterns of thinking, theories and ideas in and around geological time and the reading of fossil remains. When Tennyson's three poems are read together, this book claims, they can be seen to cohere in striking ways. One reason why the extent of the geology has not been fully appreciated is that it operates not only at the level of idea, image and metaphor but also at structural levels, and as deep structure it works powerfully but largely imperceptibly. Just as the earth itself "provides the fabric on which all experience is located", geological structures and patterns of change order the internal logic of the poems, shaping the poetic expression and dictating the ways in which the drama of each poem unfolds.

---

[2] The modern history of literary criticism on Tennyson and geology begins with Eleanor Bustin Mattes in 1951. Mattes established the terms for reading *In Memoriam*'s geology. However, most of the criticism that has followed Mattes's study has focused on *In Memoriam* and there is little written about how geology might figure in Tennyson's other mid-century poems. See "The Challenges of Geology to Belief in Immortality and a God of Love", in *In Memoriam*: The Way of the soul (1951), and Dennis R. Dean, *Tennyson and Geology* (1985). Also see, Milton Millhauser, *Fire and Ice: The Influence of Science on Tennyson's Poetry* (1971); "Tennyson, *Vestiges*, and the Dark Side of Science" (1969); "Tennyson's *Princess* and *Vestiges*" (1954). Walker Gibson, "Behind the Veil: A Distinction Between Poetic and Scientific Language in Tennyson, Lyell, and Darwin" (1958). Also see Basil Willey, "Tennyson" in *More Nineteenth Century Studies* (1956). Susan Gliserman, "Early Victorian Science Writers and Tennyson's *In Memoriam*: A Study in Cultural Exchange" (1975). Gliserman's article was perhaps the first to take account of the complex comingling of ideas and disparate discourses in *In Memoriam*, offering a sophisticated comparison of the divergent epistemological approaches found in the scientific and philosophical writings of Peter Mark Roget, William Whewell and Lyell, all of whom were read by Tennyson. For earlier critical accounts of Tennyson and science, see George R. Potter, "Tennyson and the Biological Theory of Mutability in Species" (1939); W.R. Rutland, "Tennyson and the Theory of Evolution" (1940).

Alongside Tennyson's poems, this book also pays close attention to Lyell's *Principles of Geology* as an understanding of Lyell's text is crucial to an appreciation of the poetics of *In Memoriam* and *Maud*. However, debates in and around the works of other writers, geologists, polymaths and comparative anatomists, such as William Whewell, Hugh Miller, William Buckland, Adam Sedgwick, Gideon Mantell, Robert Chambers, Richard Owen and Richard Chenevix Trench also figure in readings of the poems, and an awareness of Tennyson's understanding of the ideas of these writers and theorists modifies and enhances the perception of how and why Lyellian geology was so important to Tennyson's poetic thinking.[3] Chapter 2, for example, argues that Hugh Miller's much-neglected (in terms of Tennyson's writing) *The Old Red Sandstone* (1841) was seminal for Tennyson's geological thinking and that it figures significantly in *The Princess*'s validation of contemporary gender ideologies. The less certain ideological territory of *In Memoriam* and *Maud* indicates the move away from the firmly theological epistemological grounds of Miller's geology to the shifting landscape of Lyell's *Principles* and its 'uniformitarian' vision of change. Thus, *The Princess*, *In Memoriam* and *Maud* can be seen to plot an increasingly radical movement from the geological and ideological confidence of *The Princess*, to *In Memoriam*'s Lyellian expression of doubt and division, and on to the geologically initiated crisis in language and meaning that characterises *Maud*.

Tennyson's engagement with geology demonstrates the significance of geological language and ideas for the period, while it also illuminates the importance of poetic language in explicating the meanings and implications of geological discoveries for mid-nineteenth century thinking in philosophy, theology, and specifically for this study, for theories of the origin and nature of language. Thus, close attention to the poems and contemporary geological discourses not only enriches the understanding of Tennyson's work but also adds to the understanding of the significance of early geological science for the nineteenth century and beyond. Crucially, a close examination of Tennyson and geology also offers a remarkable insight into the origins of the bifurcation of literature and science and the emergence of science writing as a distinctly separate form, as the poems chart a crisis in the assumption of

---

[3]For a list of the publications in Tennyson's library (although many books Tennyson is known to have read do not appear in the library list), see Nancie Campbell, ed., *Tennyson in Lincoln* (1971).

the seamlessness of literary and scientific ways of knowing. In this, the poems are important not solely as aesthetic productions or as texts read within their contexts but as sites of experimentation and creativity where new geological thinking finds its wider significance and where new forms of knowing and experiencing are forged. While the focus of this study is mainly on the poetry, the geological texts are, of course, equally important in understanding this bifurcation. Reading the poetry and the geology together, the study argues, shows how they map a vital cultural shift in the history of the development of modern epistemologies.

There has been much exciting academic activity in the field of nineteenth-century literature and science in the last decade. In the area of poetry and science, there have been a number of seminal studies (Holmes 2009; Tate 2012; Brown 2013) and also in nineteenth-century literature and geology (O'Connor 2007; Buckland 2013).[4] Most scholarship, however, has taken an historicist approach, while Noah Heringman's nuanced study of Romantic poetry and geology, *Romantic Rocks, Aesthetic Geology* (2004), stands out as offering a theoretically inflected approach that addresses the origins of Romantic ways of knowing and persuasively argues that Romanticism and geology both "spring from a common source, landscape aesthetics".[5] The approach to Tennyson's poems and geology in this book makes use of a range of literary critical practices, combining historical contextualisation with close readings, but also drawing on modern literary theory and specifically Mikhail Bakhtin's model of dialogism. Theory, however, is not merely projected back onto the nineteenth-century texts as a way of understanding them via new paradigms. Rather, it is there to demonstrate the part that these poetic and geologic texts played in originating modern theory itself. Thus, one of the broadest arguments of this book is that a close examination of these texts adds to the understanding of the formation of

---

[4]For the most recent studies on Tennyson and Geology see, Virginia Zimmerman, "Tennyson's Fairy Tale of Science" in *Excavating Victorians* (2008); Aidan Day, "The archetype that Waits: 'Oh! that'twere possible', *In Memoriam*" in *Tennyson's Scepticism* (2005). On Tennyson and Darwin: Valerie Purton ed., *Darwin, Tennyson and Their Readers* (2013). On the history of science: L.I. Anderson and M.A. Taylor "Tennyson and the Geologists Part 1" (2005); M.A. Taylor and L.I. Anderson, "Tennyson and the Geologists Part 2" (2016), in press. On the novel and geology: Adelene Buckland's *Novel Science* (2013).

[5]Heringman (2004), xv.

modern critical thinking and specifically early twentieth-century language theory, as will be discussed.

This introductory first chapter begins by discussing some of the issues concerning reading literature and science together. It looks briefly at the 'two cultures' debate and at how the texts here studied—poetic and geological—are seminal themselves in the development of what we now think of as 'two cultures'. As the connections between geology and contemporary language theory are important for the book's larger argument, the introduction outlines the crucial concepts and debates around language that the texts were part of. Following this, it sets in place some of the ideas involved in reading 'remains'—geological or otherwise—in Tennyson's poems. Finally, as a preamble to Chap. 2's analysis of *The Princess*, Hugh Miller's *The Old Red Sandstone* (1841) is examined as Miller's popular theologically orientated geology figures prominently in *The Princess*, and an understanding of how it operates in the poem helps to bring into relief the heterodoxy of Lyell's geology.

TENNYSON, LITERATURE AND SCIENCE

Studying literature and science at a time when these categories are being founded raises many issues concerning their relationship. As Gillian Beer demonstrated (1983, 1996) the flow of ideas between literature and science is not a "one-way traffic"; literature does not "act as a mediator for a topic (science) that precedes it and that remains intact after its representation". Rather, "Scientific and literary discourses overlap" albeit "unstably". Beer draws attention to the way ideas shape and re-shape themselves as they move through discourses, and in turn, shape and re-shape human thought, suggesting, much "is to be gained from analysing the transformations that occur when ideas change creative context and encounter fresh readers".[6] Beer's ground-breaking work is still producing fresh insights and how nineteenth-century literature and science relate to each other is still a critical issue in itself.

Ralph O'Connor's extensive examination of geology, poetry and the scientific imagination, *The Earth on Show: Fossils and the Poetics of Popular Science*, 1802–1856 (2007), makes a number of critical points about how literature and science might be read together. One of his aims is

---

[6] Gillian Beer, *Open Fields* (1996), 173.

to address what he sees as a bias in literary criticism towards the literary text as the central form worthy of analysis. What he calls "the 'poet *x* and science *y*' framework", in which "Literature represents the centre and science one of several possible peripheries, supplying the raw material for poetic production". O'Connor offers the title of Dennis R. Dean's meticulous study of Tennyson's geological sources, *Tennyson and Geology*, as one example of what the *x y* framework tends to look like.[7] The present work, of course, is exactly that too—*Tennyson and Geology*—and the focus of this book *is*, mainly, the poetic texts. However, this study moves beyond the formula O'Connor discusses. The geologic texts are not seen as peripheral, nor as merely the source of poetic embellishment. Rather, one of the aims of this study is to apply to all texts, poetic and geologic, the same level of literary critical awareness. Thus, Chap. 3 scrutinises Lyell's *Principles* with the same attention to literary and rhetorical device as given to Tennyson's poems. But the study also aims to examine the ways in which the poetry interrogates the use of literary device in Lyell's geological text. Tennyson's reading of geology produced poems that were not merely passive receptacles of geological ideas, just as geological discourses were not merely sources for new metaphors. The poems test out the fitness of geological concepts, processes and patterns of change by probing the logic and intelligence of the metaphoric structuring used by those writing about geology, and in doing so they extend and develop the meaning of geology. They do not merely record particular intellectual shifts occurring in the period or document the cultural assimilation of things happening in geology. The poetry played a role in giving geology cultural meaning while it also helped explain geology to itself.

O'Connor also usefully questions the assumption that the "period between 1780 and 1820" saw a "watershed" in the relationship between literature and science "culminating in a divorce". Rather, O'Connor argues for the "glacial slowness of this metamorphosis", which, by 1820 "had hardly begun". For this reason, he argues, "we need to stop thinking in terms of 'given' dichotomies such as 'science and literary writings'", as "for at least the first two-thirds of the nineteenth century, this distinction cannot be assumed".[8] Thus, as the specific set of value

---

[7] Ralph O'Connor, *The Earth on Show* (2007), 446. O'Connor revisits these models in his article "The meaning of 'literature' and the place of modern scientific nonfiction in literature and science" *Journal of Literature and Science* 10, no. 1 (2017): 37–45.

[8] Ralph O'Connor, *The Earth on Show* (2007), 241, 243, 244, 448.

systems that created our present culture of division were not in place in the first half of the nineteenth century (the time just before Huxley and Arnold begin to shape the terms of the debate), texts of the period need to be treated from outside the divisions enshrined in the 'two cultures' debate. Referencing O'Connor, Charlotte Sleigh asserts that "the terms 'science' and 'literature' [as we think of them now] were both consciously constructed to satisfy particular cultural identities…and that it is therefore unwise to project a 'relationship' between them back onto the past".[9] One of the aims of this book, in its focus on the first half of the nineteenth century, is to avoid the notion of a 'relationship' between literature and science altogether. To see these presently distinct categories in a 'relationship' is to impose differences and oppositions on them that were not perceived to exist at the time when the texts themselves were created, and reading literature and science as in a relationship can actively impede the understanding of the literariness of science writing and the purpose and function of scientific ideas in the period. Not being in a 'relationship', literature and science did not 'separate' or 'divorce' as such. Rather entirely new forms emerged to "satisfy particular cultural identities" as Sleigh suggests, and to serve the demands of an entirely different set of ideological needs by developing clearly defined forms of writing emphasising particular characteristics. Thus, the geological and poetical texts addressed in this book cannot be properly understood in terms of being in a 'relationship', but must instead be seen as seamlessly integrated. The poems are in themselves complex poetic/geological texts, just as, for example, Hugh Miller's geological writings are literary productions. Both are literature and both are science by the standards of the period that did not perceive division. This is not to say that geology is not a distinct topic, but rather that the knowledge that geology imparts was perceived to be one part of a single unitary body of 'truth'. In other words, geology provided different types of evidence for what were assumed to be the same ubiquitous 'truths' that are equally capable of being expressed in both poetry and geology. Significantly, Tennyson's mid-century poems and the geological texts themselves play a particular part in demonstrating how and why it became necessary to construct 'literature' and 'science' as distinct categories. They register the stirrings of the movement towards the 'two cultures' divide and

---

[9] Charlotte Sleigh, *Literature and Science* (2011), 13.

offer valuable insights into the reasons why that divide became necessary in the first place.

Behind the contemporary sense of unity (as opposed to division) is an intellectual climate that assumed an inherent congruence between the structure of the moral and natural worlds. The same unchanging fundamental 'truths' of nature and human culture were presumed capable of manifesting themselves in manifold forms. Thus, poetic and geologic writings were potentially capable of expressing the same underlying truths of both the natural and moral spheres. In this way of thinking, it is not quite that ideas (poetic or geologic) move between different forms of writing but rather that their 'truths' are capable of surfacing in different forms writing; that is, if the writing itself is of a quality capable of bringing into relief the moral and natural truths thought to be immanent in language. It is for this reason that terms such as 'appropriation', 'exchange' and even the 'use of geology', as descriptors of how geology works in the poems, seem often to miss the mark. It was not quite the case that Tennyson saw himself as 'appropriating' geology to 'use' in his poems, although it looks like that from a modern perspective. Rather, Tennyson saw geology as already part of the 'medley', or the fabric of the poems' language. Thus, understanding the poetry and geology is not so much about looking for evidence of exchanges, for traces of influence or points of convergence. It is about looking for an emerging discordance between the poetry (literature) and geology (science). The questions to address to Tennyson's poems and the geological text are not so much about the relationship or the exchanges between literature and science but rather about where it might be possible to discern conflict and friction. Thus, reading Tennyson's engagement with geology is not an exercise in analysing the geology in the poetry (addressing the literature and science relationship); it is more about seeing from outside modern divisions and noticing where geologic and poetic ways of knowing begin to pull apart. Tennyson's poems mark a specific moment in the development of the divisions that set current critical values in place. They offer a significant insight into the mentality that did not see in terms of division, and that falls into crisis at the prospect of the emerging necessity for that division (as suggested in *In Memoriam* and *Maud*).

While the poetry and the geology are treated as equally capable of explicating natural and moral 'truths', this study does argue for the special status of the poetry, not as Literature per se, but rather in the contemporary sense that saw poetry as the highest form of all literary production. In this contemporary sense, the poetry was perceived to be capable of offering another—and possibly more valid—form of the same truths expounded in geology. Gregory Tate makes an important observation about the use of poetic allusion in science writings when he notes that "poetry in nineteenth-century culture retained its status as the height of artistic expression, and as the articulation of enduring emotional and spiritual truths". Examining John Tyndall's allusions to Ralph Waldo Emerson's poetry in his writings on atoms, Tate speculates on whether the use of poetic allusion is "mere eloquence, an expressive form of words that functions straightforwardly as an ornamentation or embellishment of the argument", or whether it could instead be considered "as an item of supporting evidence for that argument". A third possibility Tate entertains is that the allusion to poetry is not just "stylistic and aesthetic, but epistemological and even … theological or spiritual". For example, Tyndall "introduces a mysterious and arguably mystical element into his scientific argument. The 'law' to which his atoms conform is not wholly distinct from the providential 'order' that Emerson's poem identifies in the world's construction and operation."[10] The slightly earlier focus of this present study makes Tate's observations even more relevant. If poetic allusion in nineteenth-century science writing can be used as "supporting evidence" for the science, allusions to science in poetry can similarly be seen to lend weight to the scientific concept. If poetry is perceived to be the highest form of literary and linguistic production, it might also be the case that the rightness of a scientific idea is substantiated when its logic is capable of being sustained in aesthetic form. Thus, where a poem is judged on its wholeness, its unity, the harmony and cohesiveness of its aesthetic expression, the science will succeed on the strength of its seamless integration within the general internal harmony of poetic language and structure.

The crux of such a formulation is language itself and its perceived ability to embody 'truth'. Language is the meeting ground for such truths; it is the common denominator between all ways of knowing. The broader contemporary mentality that complemented the view of language as embodying truth, and to which Tennyson and his intellectual circle were finely tuned, was teleological. Teleology assumed that an overarching divine purpose worked in and through both nature and culture from first

---

[10]Gregory Tate, "The Uses of Poetry in Science" November 2015: https://drgregorytate.wordpress.com/2015/11/30/the-uses-of-poetry-in-victorian-science/.

causes towards final causes. First and final causes were part of a design that had both direction (unfolding in time) as well as ultimate wholeness (being already prefigured in the divine mind). Teleology presupposed an intrinsic order in the unfolding of history that the human imagination was hard-wired to perceive through language. Tennyson's tutor at Cambridge, the highly influential philosopher of science and all-round polymath, William Whewell, exemplifies the teleological position. Whewell coined the term 'consilience' to expound his concept of the 'consilience of inductions', the 'jumping together' of knowledge. The term refers to what Whewell saw as the concordance of evidence drawn from seemingly unrelated sources, a concordance that suggested the unity of all knowledge (as will be discuss more closely in Chap. 3). For Whewell, language was the fabric upon which consilience was woven, an assumption supported by Adamic theories of language that envisaged words as the medium through which God's divine providence becomes knowable to the human mind.

Tennyson's own approach to language was shared by the intellectual elite of his generation. As Donald Hair has demonstrated, at Cambridge, Tennyson and his circle of Apostles rejected the Lockean empiricist philosophy of their Cambridge education, which saw language as "the arbitrary linking of word and thing", turning "instead to the thinking that Mill was to label 'Germano-Coleridgian'"—a view of language that saw words not as the "labels of dead things but rather living powers". Language was linked with a "providential view of history" and a deep study of it stood to "gradually reveal a dynamic and divinely ordered world".[11] In its extreme form, as expounded by Tennyson's fellow Cambridge Apostle, Richard Chenevix Trench, who wrote widely on the subject of language and its development, Adamic theory represented a staunchly conservative view of words as embodying an immutable morality laid down by the ultimate divine authority of the past. Isobel Armstrong was the first to alert readers to the significance of Trench's writing for *In Memoriam*, suggesting that tensions between Lyell's geology and Trench's view of language lie behind the production of this "massive double poem".[12] In his "intensely reactionary work", *On the Study of Words* (1851), as discussed more fully in Chap. 6, Trench argued that words clothed what were divine ideas.[13]

---

[11] Donald Hair, *Tennyson's Language* (Toronto: University of Toronto Press, 1991), 4.

[12] Ibid., 256–7.

[13] Isobel Armstrong, *Victorian Poetry: Poetry, Poetics and Politics* (1993), 256.

Language was given to man by God but not in its fully formed state; "man" did not "begin the world *with names*, but *with the power of naming*". And "as each object to be named appeared before his eyes, each relation of things to one another arose before his mind".[14] Thus, from the Adamic approach, etymology was the study of the divine roots of words and its aim was to attempt to trace words back to their origin in the undivided totality of the divine idea.

For the poet who is the arbiter of words, the ability to conjure, in an Adamic sense, the right words in order to mobilise the "relation of things" that words evoked, and to make a cohesive unity of those chosen words, was a measure of the moral and intellectual rightness of the poetic sentiment and ultimately the success of the poem. Aesthetic value judgements thus become implicated in the authenticity of 'truth', and they are equally applicable to geologic texts as to poetic texts, as where language is perceived to embody natural and moral truth, it is in the coherence and harmony of its organisation that the rightness of the geologic or the poetic idea materialises itself. Tennyson's poetics hinged on an Adamic assumption that in his use of a word he reached back to re-animate divine ideas. Equally, however, expressing the truths of nature in language requires a deft ability to summon up the truths of creation that are already "laid up" in words (*Words*, 1). A poem's success is a mark of its manifestation of truth, and where the poem alludes to natural laws, its success confirms the integrity of the scientific concept, bestowing upon it an authenticity that it might otherwise lack.

An example of how this works can be seen in Chap. 2's reading of Tennyson's *The Princess* in conjunction with Hugh Miller's geology. Miller's own use of literary and cultural allusion in his geological texts indicates the significance of aesthetic structures for the authenticity of his geology's claim to truth. Miller, also writing from an Adamic perception of language, knew the power of poetic expression, not merely as a rhetorical tool or as a means to beautify his own writing, but as, in Tate's words, "supporting evidence", and he used it to great effect in his own geological writing. However, allusions to Miller's geology in *The Princess* demonstrate how Miller's text was not merely a source from which to lift images, metaphors or patterns designed to embellish the poetry or enhance its imaginative play. Rather, the poem supplies another

---

[14] Richard Chenevix Trench, *On the Study of Words* (1851): 15, 16. Hereafter cited parenthetically as *Words*.

type of evidence corroborating Miller's general geological vision—a type of evidence that, if anything, lends the geology added significance, as it is validated in the assumptions that poetry is the most exalted medium through which moral and natural truths are expressed. And in turn, the Millerian geology of *The Princess* is persuasive corroborating evidence supporting, as Chap. 2 argues, a truth about gender relations that is meant to be understood as proven by the fact that it is equally verifiable and expressible both in geologic and poetic form.

Where the ideological rhetoric of *The Princess* is in harmony with Miller's geology, *In Memoriam* and *Maud* represent Tennyson's experimentation with an entirely different Lyellian geology. Lyell's provocative passages on the passing of time, his comparison of human and geological time and his mammoth accumulation of evidence for the long-term radical effects of gradual, hardly perceptible processes of geological change, made *Principles* one of the most sophisticated and persuasive geological texts of its era. The elegance and rigour with which Lyell argued for an expansion into deep time has been read as a vital element contributing to a mid-century crisis in religious faith, and, more specifically, as a primary source for *In Memoriam*'s anxiety and religious doubt.[15] However, Lyell's geology had a much broader and deeper meaning for *In Memoriam* and *Maud* that hinged on his anti-teleological science. In its geology, its style, its accumulated evidence and its methodology, *Principles* was carefully designed to put in place a new set of paradigms for the study of geology—*Principles* claimed, after all, to offer the *principles of geology*. At the centre of Lyell's new principles was the methodology, and for Lyell, an immutable law of geological change, described by the term 'uniformitarianism': the position that argued that observation of present geological change explained all past geological change, and that the rates and types of geological change had not varied throughout time. The uniformitarian position has been summed up in the seemingly innocuous and yet highly provocative axiom, 'the present is the key to the past'; in other words, all that is needed to understand the workings of nature and the state of the past is rigorous attention to present process.

---

[15] Michael Tomko has offered a re-reading of *In Memoriam* and the *Principles of Geology* as texts that similarly work to revise "Paleyan natural theology into a dynamic spiritualism." See "Varieties of Geological Experience: Religion, Body, and Spirit in Tennyson's *In Memoriam* and Lyell's *Principles of Geology*" (2004).

Although *The Princess* was written while Tennyson simultaneously worked on parts of *In Memoriam*, Tennyson followed the logic of Lyell's geology for the latter poem. The result was an entirely different poetics that worked against Adamic and teleological epistemologies. Where Trench traced the divine origin of words looking back to the authority of the past and arguing that "the present is only intelligible in the light of the past", Lyell's prioritising of the present turned that logic upside-down (*Words*, 5). Where Adamic theory expressed a reactionary concern for the authority of received truths, Lyell subtly but radically destabilised Adamic authority, postulating instead entirely different grounds for the basis of 'truth'. Compounding the destabilisation of Adamic foundations, Lyell was also obliged (for his uniformitarian principles to work) to circumvent the powerful anthropocentric pull of teleological narratives, as Chap. 3 explains. To this end, he instituted a 'strategy of division' which insisted that the natural laws that govern geological agency find no analogy in the sphere of human morality. There was, in other words, one law for the material operations of geology but other unempirical, numinous laws for the unfolding of moral truths which had no bearing on the physical operations or constitution of the earth. Thus, Lyell's geology distinguished different types of truths requiring different epistemological groundings.

The radical potential of Lyell's 'strategy of division' was quickly discerned by his most perspicacious opponent, William Whewell.[16] Whewell readily saw how Lyell's insistence that geological knowledge could have no moral, metaphysical or theological bearing flew in the face of ideas of consilience. It hit at the foundations of Whewell's vision of one monolithic way of knowing and unifying the natural and moral spheres. Tennyson, well read in both Lyell and Whewell, also understood that *Principles* expounded very different sets of truths from those typically associated with teleology. Testing out Lyell's new epistemological grounding, *In Memoriam* and *Maud* are both experiments in a uniformitarian poetics. Where *The Princess* unproblematically aligns itself with Miller's teleological geology in order to expound what are seen as natural truths about gender, *In Memoriam* and *Maud* operate across Lyell's division, and the result is a radical destabilising of the poetic plane. Each poem interrogates the new epistemological landscape Lyell

---

[16]As well as being Tennyson's tutor at Cambridge, Tennyson's library contains the three volumes of Whewell's *History of the Inductive Sciences* (1837).

initiated—a landscape of disciplinary division now so familiar that the dynamic role Tennyson's poetics played in the performance of those divisions has often been missed. The poems are both implicated in the slow emergence of disciplinary division, and specifically concern themselves with that division and its consequences. Thus the crises enacted in *In Memoriam* and *Maud*, which will be the subject of subsequent chapters, came about because Tennyson's conviction that language itself embodies fundamental truths was sorely undermined when he turned to Lyell for his poetic structure.

The broad rationale for this study's reading of Tennyson's poems and geology is the proposition that they preside over a crisis in mid-century thinking about language initiated by Lyellian geology. The poems not only offer remarkable insights into the nature of that crisis and its implications for progress into modernity, they helped to forge that progress by materialising the shape of the future in their experimental uniformitarian poetics. Tennyson was not just writing *through* a time in which the language of fundamental truths was fragmenting into a multiplicity of relative values. In his response to Lyell, he is *writing that time into being*. And in turn, a sense of the failure of language to embody the truths of nature was the origin of the cultural phenomenon we associate with the linguistic turn of the twentieth century. The point of studying Tennyson and geology then is not to look for crossovers or exchanges between literature and science as such, but rather to look for the breaks, the signs of discord, and what might be called the dialogic clash of multiple truths—the places where, in other words, language itself becomes the subject. Where *The Princess* exemplifies the seamlessness of literature and science at a time when fundamental truths are seen to be potentially available in many linguistic forms, Hugh Miller's and Tennyson's texts are mutually supportive. *In Memoriam* and *Maud*, however, register a disintegration of unitary meaning and the beginning of the end of teleological authority with the emergence of new epistemologies capable of authenticating different types of truths from those associated with the human moral or intellectual sphere. Lyell's uniformitarian geology posed a fatal challenge to the presumption of the embeddedness of meaning in language and Adamic notions of the authority of the past. Thus, performing the crisis that uniformitarianism initiates, Tennyson's major mid-century poems demonstrate how the genres of science themselves emerged by default from the failure of language to sustain its immanent meaning.

Fossil Poetry

The study of geology is in many ways the study of death, not only in terms of fossil remains (the study of which eventually became its own branch of science as palaeontology) but also because, as it turns out, so much of the material earth is a memorial to lost life transformed from organic remains into inorganic substance. Organic remains are not merely inhumed into the earth, they are transmogrified into inorganic material in processes of petrifaction and mineralisation. The dead appear to be passively received into the ground, and yet they are put to work in slow but dynamic processes that incorporate them into the earth's geological structure, transforming them into the coal, stone or marble that accrues in the abysmal depths of geological time. This bleak knowledge taken from geology fuelled the Victorians' preoccupation with the materiality of death. Geology and comparative anatomy thus contributed to the Victorian fascination with the material remains of the dead and their meaning and relationship to the living. As Deborah Lutz suggests, the "Victorians expressed a desire to touch the death that surrounds life, to feel its presence as material, vital." The development of comparative anatomy as a means of re-animating fossil remains also fed into the Victorian obsession with memorial relics and suggests how "death, and the body itself" can become "a starting place for stories rather than their annihilation".[17] Fossil remains and the stories they engendered seemed to imply that on some level the past could be retained or even re-animated in the present. While Tennyson's generation expressed their fascination with the possibility of recovering the narratives seemingly embodied in remains, analogies between fossils and language appear regularly in the work of geologists. William Buckland, the foremost geologist and comparative anatomist of the first half of the nineteenth century, saw fossil remains as offering the "great master key whereby we may unlock the secret history of the earth". Fossils were "documents which contain the evidences of revolutions and catastrophes, long antecedent to the creation of the human race; they open the book of nature, and swell the volumes of science".[18] The celebrated comparative anatomist Richard Owen also saw in comparative anatomy the means to construct

---

[17] Deborah Lutz, *Relics of Death in Victorian Literature and Culture* (2015), 127.

[18] Buckland, *Geology and Mineralogy* (1836), I, 128.

narratives, to "restore and reconstruct ... species that have been blotted out of the book of life".[19]

In a similar vein, Richard Chenevix Trench began his published collection of lectures, *On The Study of Words* alluding to Emerson's notion of language as 'fossil poetry', by which, Trench suggests, Emerson "evidently means that just as in some fossil, curious and beautiful shapes of vegetable or animal life" are preserved, being "permanently bound up with the stone ... so in words are beautiful thoughts and images, the imagination and feeling of past ages, of men long since in their graves ... preserved and made safe forever" (*Words*, 4–5). Thus, Trench writes, language is the "amber in which a thousand precious and subtle thoughts have been safely embedded and preserved" (*Words*, 23). As the fossil preserves the past, each word contains its past meanings and associations. Words, like fossils, also partake of the whole, as just as fossil fragments in the hands of the comparative anatomist can be re-articulated to make a greater whole, the meaning "laid up" in words shows them to be part of larger families of words that altogether unfold a larger meaning, one that (for Trench, at least) manifests the web of divine creation. The skill of the comparative anatomist is similar to that of the philologist. Where the comparative anatomist wielded the seemingly miraculous power to re-articulate the once living body, Trench, as Megan Perigoe Stitt suggests, made the philologist "a sort of hero who rescued the feelings and ideas of the past".[20] As fossil fragments could be brought together to make a whole, so the correct and sagacious use of words could re-animate what for Trench was the original 'divine idea', as words map the related parts of an already extant whole.

The philologist was also the moral arbiter as his task was no less than to accurately trace the origins of words back to God's ordained system of morality. Thus, extending Emerson's analogy, Trench asserted that language might be thought of not merely as fossil poetry, but as "fossil ethics, or fossil history", as words "quite as often ... embody facts of history, or convictions of the moral common sense, as of the imagination or passion of men" (*Words*, 5). Like the philologist, the poet also wields the moral power invested in words, and as already suggested, the assumed Adamic power of language is implicated in aesthetic judgment.

---

[19] Richard Owen, *Lectures on the Comparative Anatomy and Physiology of the Vertebrate Animals*, 2 vols. (1846), II, 3.

[20] Megan Perigoe Stitt, *Metaphors of Change* (1998), 2.

Tennyson's *The Princess*, which, as I will argue, invests heavily in the authority of the past in terms of fossil evidence, offers an example of how the seeds of destruction for the poem's rhetoric and for Adamic theory lie, ironically, in this very investment in Trench's Adamicism. For Trench (often tetchy in his dealings with Tennyson), *The Princess*'s designation as 'medley', over and against the use of the word 'mellay' in the poem, indicates a misuse of the meaning embodied in language. Where 'Mellay' is used to describe Arac in the throes of battle; "He rode the mellay, lord of the ringing lists" (V, 491), the speaker of the Prologue introduces the eclectic scene at the Mechanics' Festival (and presumably the several parts of the story) declaring "This were a medley" (230). As in Trench's reference to fossil poetry, the expectation is that the poem's medley of words and parts can be coherently assembled to rearticulate natural, pre-existent truths, in this case truths concerning gender. In his reference to a 'medley' of parts, Tennyson treats words like fossil fragment that can be brought together to re-animate their living powers and the truths they embody. In this, Tennyson takes his cue from Gideon Mantell's description of the comparative anatomist who uses his skill to order a "confused medley of bones" into a coherent whole. For Mantell, it takes the comparative anatomist knowledge of the "co-relation of structure in organized beings" to "decipher the hand-writing on the rock" and to reassemble the 'medley' of fragments into a complete whole true to life [21]. If the comparative anatomist failed to correctly order the fossil fragments before him, he risked creating a 'monster'; a thing untrue to nature. Similarly, the poet's organisation of words enables him to excite the latent power invested in them, and the measure of a poet is in his ability to order words according to their 'true' moral context and not to pervert their meaning in the loose and ill-informed organisation of language.

'Medley', strangely enough, was also one of Trench's contested words. In his *A Select Glossary of English Words Used Formerly in Senses Different From Their Present* (1859), Trench bemoans the loss of the word's original meaning as 'conflict' and 'combat in battle'. Specifically referencing *The Princess*, his entry for 'medley' reads:

---

[21] Mantell, *Wonders of Geology Or, A Familiar Exposition of Geological Phenomena* (1838), I, 127, 128.

It is plain from the frequent use of the French 'mêlée' in the description of battles that we feel the want of a parallel English word. There have been attempts, though hardly successful ones, to naturalize 'mêlée', and as volée' has become in English 'volley', that so 'mêlée' should become 'melley'. Perhaps, as Tennyson has sanctioned these, employing 'mellay' in his *Princess*, they may now succeed. But there would have been no need for this, nor yet the borrowing of a foreign word, if 'medley' had been allowed to keep this more passionate use, which once it possessed.[22]

The subtle implication here is that Tennyson's unnecessary employment of 'mellay' to mean conflict in battle foregrounds his misuse of the word 'medley' as a designation for *The Princess*'s potential unity. Trench's criticism obliquely points to what he sees as Tennyson's fallibility as a poet. However, for a poem that invests so heavily in the authority of the past in both its use of geology and in its parabolic form, the confusion is still more telling. It demonstrates how Adamic theory is apt to undo itself. Where for Tennyson, the 'medley' anticipates an inherent unity; for Trench, strictly speaking, it is struggle—it is the clash of warring factions that cannot be harmonised, and this nuance of definition (in Adamic terms at least) skews the whole poem's rhetoric. Trench might well have agreed with the contemporary reviewer of *The Princess* for whom the medley of "Lecture rooms and chivalric lists, modern pedantry and ancient romance, [...were] antagonisms which no art can reconcile"[23]. The poem risks being a conglomerate monster with no 'truth' in life. More significantly, however, what Trench foregrounds unintentionally in his criticism of Tennyson's use of 'mellay' is the instability of language itself—language's ambiguity; its protean nature and its dialogic quality. *The Princess*'s medley aims for a grand unity that, because of the scale of its ambition, easily flips over into dialogic chaos if any one part of it fails to tally up with the central truths the poem aims to manifest. This clash of meaning is a heteroglossic Babel; the very thing Trench sought in his writings to curtail, and the very thing that Tennyson confronts in *Maud* when he loses his grip on Adamic theory in his experimentation with Lyell's anti-teleological, present-focused uniformitarian geology.

---

[22] Trench, 131.

[23] J. W. Marston, unsigned review, The *Athenaeum*, 21 (January 1, 1848), 6–8. Quoted in Shannon, *Tennyson and the Reviewers* (1967), 98.

BETWEEN A ROCK AND A HARD PLACE

Trench's allusion to fossil poetry also draws attention to the narrative potential of fossils. Fossils could be brought together to reconstruct more than mere shape. Hugh Miller, for example, described how for Cuvier, "The condyle of a jaw became in his hands a key to the character of the original possessor." From mere bone fragments, Cuvier could "read a curious history of habits and instincts" (*ORS*, 147). As the comparative anatomist revealed the 'character', 'habits' and 'instincts' of past life, the philologist revealed the 'moral common sense', 'imagination' and the 'passions' of the past, as words "often contain a witness for great moral truths" (*Words*, 9). The ability of the philologist and the comparative anatomist to extrapolate from words or remains offered them a strange kind of fascinating but teasingly unsatisfactory ability to reclaim the past. Tennyson used words to animate a web of associations, and similarly fossil remains, as fragments to be reassembled or recovered, or as objects to be read, are frequently the source of narrative construction in his poetics. The speaker of *The Princess*, for example, attempts to order the relics, fossils and miscellaneous objects, as well as voices, songs and lyrics into a narrative whole, as critics have noted.[24] However, when faith in the authority of the past is undermined, the poetic confrontation with remains has more disturbing consequences, as in *In Memoriam*, where the speaker agonises over his inability to reconstruct Hallam's "loved remains" (IX, 3), and in *Maud*, where the narrative unfolds via the speaker's deranged readings of disparate geological and memorial remains.

Before exploring the importance of reading remains, the use of the term 'remains' requires some justification. Virginia Zimmerman, in her extensive examination of Victorian geology and archaeology, uses the term 'trace' in reference to Jacques Derrida's *Of Grammatology* to encompass both the geological fossil and the archaeological artefact. However, the term 'remains' is preferred in this analysis of Tennyson's poetry, as while Tennyson's tropology is largely geological, the poems embrace a more general notion of remains that includes not only inanimate geological objects—rocks and stone—but also other fragments,

---

[24]For critical comment on *The Princess* and geology and evolutionary theory respectively, see Zimmerman, Chap. 3, Rebecca Stott, "Tennyson's Drift: Evolution in *The Princess*," in Purton ed., *Darwin, Tennyson* (2013).

such as the fossil "rude bones" (III, 279) of the prehistoric animal in
The Princess, as well as the more recent remains of the dead—the body
of Hallam in *In Memoriam*, the empty shell in *Maud* and the memorial ring made from his mother's hair that Maud's brother wears.[25] The
term also emphasises the memorial connotations of Tennyson's engagement with the fragments of the past and helps to figure the paradox that
such objects embody: that remains are dead but that they nevertheless
'remain'. This contradiction is often at the root of Tennyson's play on
geological objects and more recent human remains.

Zimmerman draws on Paul Ricoeur's *Time and Narrative* (1985)
to explore the "relationship between material things and time". As she
points out, the trace "offer[s] a measure not only of time's passage but
also its peculiar stasis". Thus, "traces, for all their connection to the past,
are emphatically of the present".[26] Remains of the past also rupture our
sense of history and its temporal flow. Michael Shanks and Christopher
Tilley explain the contradictory nature of our confrontation with the
past:

> The past is conceived as completed. It is in grammatical terms 'perfect,'
> a present state resulting from an action or an event in the past which is
> over and done. This 'perfected' past is opposed to the flow of the ongo-
> ing, incompleted, 'imperfect' present. Although the past is completed and
> gone, it is nevertheless physically present with us in its material traces.[27]

The fossil, relic or memorial fragment presents the observer with the
anomaly of a past which is dead and complete and yet remains in the present somehow (seemingly, at least) outside time, and Tennyson's speakers in *In Memoriam* and *Maud* are frequently figured confronting this
incongruity. The nub of this problem is that the apprehension of remains
in the present foregrounds both the observer's own incompleteness and,
perhaps more troublingly, that this incompleteness is a condition of consciousness. Remains, signifying that which is complete, past and over and
done with, also speak tangibly of a closure that can never be achieved

---

[25] All references to *The Princess* and *In Memoriam* are taken from *The Poems of Tennyson*,
Christopher Ricks ed. (1969). All references to *Maud* are taken from *Tennyson's Maud:
The definitive Edition*, Susan Shatto ed. (1986).

[26] Zimmerman, *Excavating Victorians* (2008), 8.

[27] Quoted in Ibid., 8.

in consciousness—that can never be achieved by the viewer of remains. Remains are thus unwelcome reminders of the paradox of consciousness—that it can never enact its own closure—a paradox that lies behind *In Memoriam*'s speaker's inability to move forward, stuck as he is in a present in which Hallam no longer exists. It is also a feature of *Maud*'s nightmarish vision in which the speaker imagines himself dead and buried like a fossil in the strata beneath the city streets. In *Maud*, unable to envisage the future self as dead remains, the speaker envisages instead the equally impossible alternative of being dead, but 'remaining', and remaining conscious. Thus, remains powerfully confront the individual with a dead past still present, reminding them of that which is impossible to envisage, their own objectified self, their ultimate assimilation into a material world that is at once the stuff of quotidian life—all we know and the very ground we walk on—yet is also intrinsically alien, unfathomable and unbearable to behold in consciousness.

Remains serve a variety of different purposes in Tennyson's poetic experimentation. They emphasise how all living things, and even entire civilisations, inevitably become the remains that a future age will contemplate. Tennyson was fully aware of geology's message about the future from his reading of *Principles*. Lyell demonstrated how the fate of all life is to be subsumed into the earth. Lyell wrote extensively on processes of fossilisation in *Principles*, as Chap. 5 examines. The animated present, Lyell shows, will inevitably solidify and be fixed into inanimate remains for future generations to contemplate. Remains, it seems, are tied to the past, but they are also prophetic of the future, as while they signify the past, they are also undeniable signifiers of the inevitability of our future state of being, or non-being. In this way, remains point both backwards and forwards, and the contemplation of remains produces a moment in which the co-ordinates of past and future intersect to sharpen the present. Here, in this moment, the past and the future are in fact more stable than the present—the past completed, the future (as predicted by remains) the only certainty—the two meet in a seemingly far less stable now, in other words, the consciousness of the observer. Consciousness is plotted by remains; it exists somewhere between a past characterised by its inaccessibility and the future certainty of the death of the subject—between the past embodied in the material relic, and the inescapable knowledge of our own future as inanimate relic. Thus, consciousness exists unstably between, what we might call, a rock and a hard place. Yet, this state of instability is the stuff of life; it provides both a heightened and exhilarating experience of consciousness (the conscious awareness of not being remains) and a realisation that the 'self'—which is only contained within this fragile moment of self-realisation—is made aware of its condition in the present via the contemplation of remains.

The focus on remains also demonstrates a mode of experimental thinking in Tennyson's mid-century poems perhaps best understood via the theoretical writings of Mikhail Bakhtin. The realisation of the self that remains offer is one that abides by the conception of the self as subject in language. For example, in *Maud*, the speaker's self-aware-ness comes into being in the processive interplay between speaker and remains, subject and object. The object, in other words, brings the sub-ject into self-awareness in the process of mediation, in the recognition of division that defines the self (the speaker) as *not* the object.[28] Remains not only have narrative potential as fragments that, in the manner of comparative anatomy, can be read and reconstructed in the present, they also foreground how narrative constructs the observer. Reading remains, then, is not only a way of writing the object back into being, it is also a way of writing the self into being, of seeing the self in the unfinalis-able present of consciousness. It is in the moment of confrontation with remains that the self's awareness of consciousness, as entirely existent in the present and intrinsically unfinalisable, is heightened. The speaker of *In Memoriam*, for example, denied Hallam's remains, loses his narrative sense of self; he is a present 'I' caught in a non-progressive uniformitar-ian landscape in which linearity is lost.

Remains gesture towards the past yet foreground the present as the site of lived being, and in this they exhibit a dialogic quality. In their potential to elicit narratives from the observer, they might also be seen as chronotopic objects. The chronotopic object deprives the observer of their transhistorical perspective by foregrounding how the narratives that objects elicit from the viewer are not absolute but time/space spe-cific. Remains fix a complex set of temporal and spatial co-ordinates in the moment they are confronted, but with each new moment of con-templation, temporal and spatial co-ordinates are remade. The 'fixing' of co-ordinates belies the instability of meaning that infinite interpretive possibilities offer; the vital emphasis here is on the present and on how meaning is created in a dynamic and shifting present and not embodied

---

[28] See particularly E. Warwick Slinn, Chap. 2, "Consciousness as Writing," and Chap. 3, "Absence and Desire in *Maud*," in *The Discourse of Self in Victorian Poetry* (1991).

in the solid object of the past. Meaning is emphatically not immanent in remains; radically, it is the product of consciousness in the present, and as such it operates to undermine the authority of the past encoded in the teleological narratives that claim to know one monolithic version of both the past and future. The focus on the present as the site of dialogic possibility draws attention back to Lyell's uniformitarian approach. As the present is the key to the past, so, in turn, remains sharpen awareness in and of the present. The narratives that remains elicit are not implicit in remains, but rather are a projection of present consciousness without which they have no intrinsic meaning at all. Similarly, words—the fossil poetry of Trench's Adamic theory—lose their fixed and immanent meaning in the uniformitarian prioritising of present process over an unknowable past. The loss of faith in the ability of remains to embody the past is also a loss of faith in the concept of Adamic language, as words, like remains, now appear to have no intrinsic meaning, they merely reflect back what is projected onto them. Where the present is the key to the past, it is in the utterance that words find their meaning; they emerge in the dialogic exchange or the microdialogue of consciousness, they take on a provisional meaning according to the time/space co-ordinates of the moment, outside of which they fall back into meaninglessness.

The present, however, is not only the key to the past. In Lyell's uniformitarianism, where the past is made to conform to the present, present processes of change are extrapolated in order to envisage the future. As Paolo Rossi suggests, uniformity "permit[s] us to foresee the continuity of future changes".[29] In Lyell's striking passages that predict the embedding and fossilisation of whole cities over the course of time, he extrapolates processes in the present to envisage a future in which his readers and subsequently all humanity become remains and are incorporated into the earth's strata. His ability to predict the future comes from the observation of the present and from the uniformitarian postulation that the present is not only the key to the past but also the key to the future.

A word needs to be said in terms of the use of Bakhtin with poetry. As Isobel Armstrong notes, Bakhtin did not believe "that poetry could generate dialogic structures or that poetic texts could participate in struggle", yet, she suggests, "Manifestly ... the Victorian double poem

---

[29] Paolo Rossi, *The Dark Abyss of Time: The History of the Earth and the History of Nations from Hooke to Vico*, 1979, trans. Lydia G. Cochrane (1984), 116.

generates the drama of contending principles".[30] Victorian poetry can certainly benefit from Bakhtinian scrutiny. However, in this examination of Tennyson's poetics, the poetry is not read as dialogic but rather as setting up the conditions, via geological patterns of thinking, and specifically Lyell's uniformitarianism, in which the speakers of both *In Memoriam* and *Maud* find themselves confronted with the possibility of dialogism as a social reality and as a condition of language and being. This is very different from reading the poems themselves as producing an aesthetic representation of dialogic struggle. This reading does not claim that any of the poems are dialogic; rather, it argues that they come into close proximity with what would later be theorised as a dialogic perception of the self in language. It was in this dialogic perception of language that Tennyson's poetics moved away from Adamic and teleological ways of knowing. And Tennyson's dialogic perception, I argue, was a direct consequence of his experimentation with Lyell's geology. At its extreme, Tennyson's uniformitarian poetics posits his speakers at the very edge of meaning. Constructed by the remains they contemplate, and at times aware that their narrative constructions are mere projections of meaning onto dead remains, these speakers exist in a geologic and uniformitarian present that dissolves the authority of the past and that leaves them disconnected from the linearity of their personal narratives and of history itself. Tennyson's uniformitarian poetics pre-empts the post-structuralist and postmodernist linguistic turn, materialising the instability that would become the mark of twentieth-century linguistics. Thus, quite remarkably, while Tennyson's mid-century long poems are self-consciously grounded in their historical contexts, in their experimental encounter with language as dialogic, they look to the future. In them, modernity is envisaged and performed, and in this they are—like uniformitarianism itself—prophetic, positing a future hardly capable of being imagined in any other writing of the time.

HUGH MILLER AND *THE PRINCESS*

While *In Memoriam* and *Maud* are poems of crisis, both destabilised by Tennyson's Lyellian uniformitarian poetics, *The Princess* is notably different, drawing principally on Hugh Miller's geology rather than Lyell's.

---

[30] Armstrong, *Victorian Poetry* (1993), 493 n. 34.

Lyall Anderson and Michael Taylor note the "surprising omission" of Miller in considerations of Tennyson's poetry written in the 1840s.[31] Tennyson seems to have felt a poetic sympathy with Miller's geology, and there is evidence in all three major poems that he had read Miller's *The Old Red Sandstone* (1841) and had been impressed with its language, its fine passages of description and its imaginative flourishes.[32] However, while *The Old Red Sandstone* made imaginative and poetic enquiries into the condition of prehistoric worlds and eloquent asides on the sublime nature of geology, the text's science was by no means as impressive as Lyell's. The comparison, of course, is unjust, as Miller did not set out to offer the 'principles' of geology as Lyell did but to educate, entertain and spiritually uplift a readership whom he addressed directly, as "working-men". As a stonemason risen to literary fame, Miller knew the value of education and knowledge. He was a model of the Victorian self-made man and of the self-help ethos of the era, featuring in Samuel Smiles's *Self-Help* (1859) as "a truly noble and independent character in the humblest condition of life".[33] His geology, however, like Trench's language theory, was firmly Adamic and teleological, and his rhetoric was in

---

[31] Anderson and Taylor, "Tennyson" (2015), 341–2. During his 1848 tour of Cornwall, for example, the young Elizabeth Rundle, records her conversation with the poet: "Then he turned to Geology ... 'Conceive,' he said, 'what an era of the world that must have been, great lizards, marshes, gigantic ferns!'... I replied how beautiful Hugh Miller's descriptions of that time are: he thought so too" (*Memoir*, I, 277). Later, in 1857, Tennyson acknowledged a gift of Miller's *The Testimony of the Rocks* (1857) from the Duchess of Argyll (*Letters ALT*, II, 178). More specifically, Christopher Ricks notes, with reference to *In Memoriam* section LVI, the "jotting by T. in H.Nbk 18 [which] acknowledges his reading of Miller's *The Old Red Sandstone*, writing the title in brackets next to his 'descriptive jottings'" (*Tennyson*, 1989, 399n.). Also see, Christopher Ricks ed., *Tennyson: A Selected Edition Incorporating the Trinity College Manuscripts* (1989), 50. Both the diary entry and the notebook are from 1848, which suggests that Tennyson had read Miller before 1848, although there is no evidence as to exactly when. Despite reading Miller and appreciating his literariness, Tennyson seems to have downplayed his interests in Miller. I think this is partly to do with Miller's didacticism and the unsophisticated pitch of Miller's address, which would have jarred with Tennyson.

[32] Dennis Dean suggests that "Tennyson was deeply affected by his delayed reading of Lyell's *Principles* and other current works on geology, among which were Charles Babbage, *The Bridgewater Treatise* (1837). Lyell's *Elements of Geology* (1838), Gideon Mantell's *Wonders of Geology* (1838), Hugh Miller's *The Old Red Sandstone* (1841), and Lyell's *Travels in North America* (1845)." (1985), 10.

[33] Samuel Smiles, *Self-Help* (1859) 98. For Miller's own account of his life see, *My Schools and Schoolmasters; or, The Story of My Education* (1854).

tune with a conservative insistence on the fixity of both species boundaries and the social order, treating both in his geological text.

The Old Red Sandstone opens with a no-nonsense address to the working classes, warning them against the narratives of political empowerment offered by Chartism:

> My advice to young working-men, desirous of bettering their circumstances, and adding to the amount of their enjoyment, is a simple one. Do not seek happiness in what is misnamed pleasure; seek it rather in what is termed study. Keep your consciences clear, your curiosity fresh, and embrace every opportunity of cultivating your minds. You will gain nothing by attending Chartist meetings. The fellows who speak nonsense with fluency at these assemblies, and deem their nonsense eloquence, are totally unable to help you or themselves; or if they do succeed in helping themselves, it will be all at your expense. (*ORS*, 1)

Again addressing the working man directly, Miller unambiguously tackles the issue of class inequality: "You are jealous of the upper classes; and perhaps it is true that, with some good, you have received much ill at their hands." But, he warns, "upper and lower classes there must be, so long as the world lasts" (*ORS*, 2). Tellingly, having asserted the inevitability of class inequality, Miller, following Lyell's lead in *Principles*, set out his objection to Jean-Baptiste Lamarck's theories of transmutation (1809)[34]:

> The ingenious foreigner ... concluded that there is a natural progression from the inferior orders of being towards the superior; and that the offspring of creatures low in the scale in the present time, may hold a much higher place in it, and belong to different and nobler species, a few thousand years hence. The descendants of the *ourang-outang*, for instance, may be employed in some future age in writing treatises on Geology ... Never yet was there a fancy so wild and extravagant. (*ORS*, 39)

Miller's progress from his political warning to working men in his introduction to his admonishment of what he ridicules as Lamarck's fanciful, wild and extravagant notions of transmutation was a deliberate attempt to manage the reception of geological ideas and to curb the production of those materialist and evolutionary theories that geological evidence

---

[34] Chapter 3 examines Lyell's refutation of Lamarck.

appeared to sanction. Like so many of his contemporaries, Miller saw hierarchical patterns to be as fundamental to social cohesion as they appeared to be to the natural world. From the order of the family unit, to class stratification and species rankings, a natural and rightful hierarchy exerted itself for the harmony of the whole, and any disruption of that order went against the divine organisation of things inscribed in nature by the "adorable Architect" of creation (*ORS*, 96).

In these concerns and in Miller's unambiguous address to working men, *The Old Red Sandstone* can be read as part of the contemporary middle-class attempt to steer an increasingly literate working-class towards what was seen as 'useful knowledge' in the hope of managing their intellectual understanding and diverting them away from speculation on the implication of new discoveries in geology (and other sciences) for religious, philosophical and political concerns. Such concerns are also registered in *The Princess* in the framing narrative's examples of harmonious class relations set against the concordant mix of educational pursuits, entertainment and the dissemination of technical and scientific knowledge. The importance of the poem's representation of class for its argument on gender has been thoroughly explicated by John Killham in his meticulous account of *The Princess* and those contemporary theories concerning women's education with which Tennyson was likely to have been familiar. Killham surveys the ideas that were propagated in France and Britain (principally by the Saint-Simonians and Owenites respectively) and establishes important connections between socialist discourses on the rights of the working classes and contemporary polemics on the position, education and rights of women in society. Such connection, Killham argues, "makes the description of a Festival of a Mechanics' Institution so appropriate to a scene also containing a party of young men and women discussing the legitimacy of women's aspirations".[35] Miller's *The Old Red Sandstone*, as Chap. 2 argues, demonstrates the important role geology played in Tennyson's thinking around class and gender inequality. Miller's direct addresses to working men, for example, marks him out as one of the "patient leaders" (58) of the Mechanics' Institute mentioned in the Prologue of *The Princess*. His measured, comprehensive writings, his explanations to "lowlier readers", and those who "may possibly be repelled" by "unfamiliar" Greek terms, make him the

---

[35] John Killham, *Tennyson and The Princess: Reflections of an Age* (1958), 65.

consummate intermediate between mechanics and geological knowledge (*ORS*, 13, 33). Couching his geological writings in an Evangelical reverence, Miller patiently 'leads' his "humble" readers, inviting them to think both imaginatively and technically, while steering them firmly towards speculation circumscribed by divine authority and away from a materialist interpretation of nature. Like the entertainments patronised by the landed gentry that divert the crowd at the Mechanics' Festival in the Prologue of *The Princess* (the "fountain", "knobs and wires and vials", the "cannon", "telescope", "electric shock", "clock-work steamer", "railway", "fire-balloon" and "telegraph" (Prologue, 54–80)), Miller's text set out to manage mechanics' exposure to geology, inspiring wonder with his imaginative passages but always bringing his readers back to the "adorable Infinite" who is the first and last cause.

The Princess was, as already mentioned, written in the same period that Tennyson was working sporadically on the lyrics of *In Memoriam*, and Rebecca Stott suggests that the poem "should be seen as being in open dialogue with its pair, *In Memoriam*".[36] The two poems, however, are opposing in their geology—an opposition that is the result of their different aims and different forms of address. Miller's text provided Tennyson with a vision of the geological past through which to frame his anti-feminist and reactionary rhetoric. While *The Princess*'s defence of conventional gender ideologies requires a stable base, which the poem finds in Miller's espousal of a fixed and harmonious natural and social hierarchy, *In Memoriam*'s expression of crisis is figured through the unstable, continuously shifting ground of Lyell's geology. The movement from Miller's geology to Lyell's produced a distinctly different poetics, one based, as already suggested, on the particular perception of geological change and geological remains that Lyell's geology encouraged. Where the uniformitarian 'I' of both *In Memoriam* and *Maud* is a subjective 'I' which appears at times to be losing (*In Memoriam*) or has lost (*Maud*) its grip on ideology, the 'I' of *The Princess*, which is a conglomeration of voices (the seven narrators of the poem's story), is the univocal voice of ideology speaking through that most ideological of forms, the parable. As Hallam Tennyson wrote of his father's choice of the parable; for the "supreme meaning and guidance of life, a parable is

---

[36] Stott, *Darwin, Tennyson and Their Readers* (2014), 32.

perhaps the teacher that can most surely enter in at all doors".[37] The 'I' of *The Princess* is supremely ideological; its message is reactionary, and crucially, it is stabilised by the weight of the past that the parable's form carries with it. The poem's parable is the construction of the male narrators who pass between them a tacit, inherited 'wisdom' about nature and gender. They fabricate the story in the manner of the "'Tale from mouth to mouth' game" that Tennyson and his "brother undergraduates" played at Cambridge: "if he 'that inherited the tale' had not attended very carefully to his predecessors, there were contradictions; and if the story were historical, occasional anachronisms" (*Memoir*, I, 253). In the poem too, it is essential to get the story straight, to match, in other words, the story with the idea, and to pass it down with its ideological substance intact. Thus, the implied improvisation of the mouth to mouth game belies a compliance to the narratives of the past. The seamlessness of the seven narrated parts attests to the importance of shaping the parable's ideology into natural, universal wisdom. The 'medley' of voices makes a harmonious whole, while the parable resists critical interrogation, claiming to merely reinstate accepted 'truths'. The parable is meant to manifest through language the ideas that already exist in the very fabric of creation, in the same way that Trench saw the "relations of things" just waiting to be brought into relief by the power of naming innate in man.

All, apart from Walter, who as Sir Walter Vivien's son represents aristocratic continuity, are nameless narrators who 'fall in' with the tale, each adopting the role of Prince, as in section V, where "he that next inherited the tale" "assumed the Prince" (16, 26). This 'assumption' universalises the masculine position and fixes the poem's thinking on gender, gesturing towards assumptions of unity outside the text as well. And, just as all the narrators' voices merge in the Prince, there is a merging of voices in Hallam Tennyson's treatment of the poem in this memoir of his father. Often it is difficult to know who is speaking in the *Memoir*: the son moves from talking about his father, transcribing notes left by his father, to reporting his father's speech and the speech of others about his father. In fact, the *Memoir*'s chapter on *The Princess*, its inception and critical reception, is as much a medley of male voices as the story of Ida itself—as if the narrative message seeps in and out of the fictive

---

[37] Hallam Tennyson, *Tennyson: A Memoir by his Son*, 2 vols. (1897), 1, 249. Hereafter cited parenthetically as *Memoir* by volume and page number.

form in the reproduction and reinforcement of the same ideologies. Like the poem, this conflation constitutes a male narrative about the 'woman question'; it is a narrative inherited and passed down via a univocal 'I', and through the safe hands of the cultural arbiters of the day. Through this univocal voice, *The Princess* reaffirms what it takes to be the naturalness of gender inequalities. However, as already suggested, by forming its gender rhetoric around an idiosyncratic reading of Miller's geology, it also reaffirms the geology in the poetic cohesiveness of its gender message. Operating to expound the same truths (truths that in Adamic terms are already immanent in language), geology and gender in *The Princess* can be seen as mutually supportive, each supplying evidence of the other's intrinsic truths.

Chapter 2 explores in depth how geology and gender work in *The Princess*. Possibly more than any other of Tennyson's major poems, because of its claim to engage with contemporary debates, *The Princess* exposes the ambivalence with which Tennyson viewed so many current issues. Such ambivalence is demonstrated in the conflict between his willingness to explore and experiment with forms and themes that interrogate social conventions and his dread of the erosion of class and gender divisions that such experimentation often envisages. The chapter examines how geology is used to shore up gender ideologies and how the poem's reactionary use of Miller's text suggests a confidence in geology's ability to express absolute truths which is wholly lacking in the *In Memoriam* and *Maud*. Chapter 2 demonstrates how *The Princess*, often read as sympathetic to the feminist cause, figures instead a reactionary gender message that is drawn from Miller's understanding of fossil remains and the natural hierarchy they indicate. It explores Miller's geological vision of serial creation and extinction to show how his teleologically orientated geology is redirected in the poem in order to correct Ida's aberrant feminism and reinstate a gender hierarchy in tune with conservative mid-nineteenth century ideologies. Thus, the poem's gender message is figured through an internal logic based on the evidence of fossil remains. The poem's message is: if Ida's desire is to leave a feminist legacy (her "sandy footprint harden[ed] into stone") then she should look to the lessons of geology to see what such fossil imprints memorialise.

Chapter 3 focuses on Lyell's *Principles of Geology* along with the work of William Whewell, both of whom were read by Tennyson. The main focus is *Principles* and its uniformitarianism, as understanding this concept is seminal to grasping the uniformitarian poetics of *In Memoriam* and *Maud*. The chapter looks specifically at Whewell's characterisation of Lyell's scientific approach as 'uniformitarian' in opposition to geological 'catastrophism'. It also examines the nuances of Lyell's rhetoric and Whewell's critique of *Principles*. Whewell's fascinating writings have received little literary critical attention; here, however, they are treated alongside Lyell's text in order to analyse the rhetorical strategies employed by both men and the importance of this dialogue for both understanding the history of geology and Tennyson's poetics. Examining Lyell's problematic insistence on a division between the natural world and the sphere of human intellect and morality, the chapter looks at how that this complex strategy of division undermined teleology and also distorted Lyell's science, and at how Whewell responded to the effects of Lyell's division, seeing it as a challenge to his own perception of the cohesive nature of the historical sciences as expounded in his concept of consilience. Whewell's criticism of Lyell helps to expose the fault lines in Lyell's theorising, and a knowledge of the tensions between Lyell and Whewell is vital to understanding the conflicts that energised Tennyson's *In Memoriam* and *Maud*.

Chapter 4 turns attention to *In Memoriam*. Where Miller's geology was fine-tuned to endorse gender assumptions in *The Princess*, in *In Memoriam*, Lyell's uniformitarian laws subject the poem to a different set of geological principles. Lyell's geology worked to erode teleological assumptions about the centrality of humanity in creation and the presumed authority of the past. It is in the differences between *The Princess* and *In Memoriam* that the radical nature of Tennyson's geological thinking becomes clear. Tennyson's willingness to move from the privileged centre ground of ideology as given in *The Princess*'s Millerian geology, into the marginal landscape of *In Memoriam*'s Lyellian form, suggests an unwavering commitment to test out the implications of new geological ways of perceiving relations between nature and the human world. The tradition in Tennyson scholarship has been to read Lyell's geology as responsible for *In Memoriam*'s most profound expressions of religious doubt. The 'geological' sections (50–56, for example) are routinely cited as evidence of Tennyson's engagement with Lyellian geology. However, Chap. 4 argues that what has been missing in the geological readings of this most celebrated elegy is an analysis of Tennyson's application of Lyell's uniformitarian *principles* as organising principles in poetic terms; it was this innovation that made *In Memoriam* a uniformitarian poem on numerous formal levels. The chapter argues that *In Memoriam* is structured on four basic principles postulated in Lyell's *Principles*: Lyell's strategy of division; the principle of 'displacement' and the repetition that displacement enacts; the principle of non-progressive change; and finally, Lyell's uniformitarian insistence that present processes have been uniform in type and rate throughout geological time. Employing uniformitarian laws as poetic ordering principles results in a uniformitarian poetics in which the speaking 'I' becomes a uniformitarian 'I' disconnected from the past and acutely conscious of itself in the present. The tensions between theme and form, recovery and grief, and past and present are intrinsically geologic and expand into wider cultural tensions between, on the one hand, authority, history and teleology, and on the other, the radical potential of uniformitarian change in the present. The uncertain 'I' of the poem is one that is in process; it is a self that seems unable to be fully formed and unable to realise itself in ideological form because it is bound to a Lyellian material landscape which has been emptied out of direction and purpose. One of Tennyson's great achievements is to show where Lyell's uniformitarian vision leads. He demonstrates a sophisticated comprehension of both the scientific and cultural implications of Lyell's *Principles* virtually unrivalled in its time in any type of text. The full critical implication of Lyell's uniformitarianism, however, is not worked through in *In Memoriam*, rather it is left to *Maud* to follow Lyell's geology to its logical conclusion. This is partly to do with form, as *Maud*'s formal designation as a monodrama allows for the fictional character's breakdown, while such a dissolution of self, imminent as it might appear at points for the speaker of *In Memoriam*, would be fatal to the elegiac expression.

Tennyson saw the anarchic power of the uniformitarian gaze, and as he progresses from *In Memoriam* to *Maud*, the radical implications of Lyell's uniformitarianism, particularly for meaning in language, become increasing clear culminating in *Maud*'s dialogic crisis which is the subject of both Chaps. 5 and 6. Such space is given to the poem for a number of reasons. In part, it deserves the attention because it has rarely been examined in terms of geology. The readings here set out to demonstrate that *Maud* represents the climax of Tennyson's engagement with Lyell's uniformitarianism. Chapter 5 explores Tennyson's reading of geological processes of fossilisation and the embedding of organic remains in inorganic structures. Much of Tennyson's understanding of these processes came from his reading of Lyell, but other sources are also explored. His wider readings of works by William Buckland, Gideon Mantell and Richard Owen also colour his interpretations of Lyell. The focus in this chapter is *Maud*'s articulation of remains (geological, fossil, biological) in the fashion of early nineteenth-century comparative anatomy. The chapter examines the ways in which procedures in comparative anatomy offered parallels with the construction of narratives and the deciphering of meaning in language. Via contextual evidence and close reading, the chapter seeks to show the vital significance of remains for the speaker's construction of his own self. As so often in Tennyson's uniformitarian poetics, the self's experience is expanded into collective human experience, and the speaker's crisis, which is bound up in the poem with issues of class, property and economics, is broadened out into a social crisis in which the Mammonism of the age is encoded in images of petrifaction as the living characters of the poem appear to transmogrify into geological remains in the present and in human time.

Chapter 6 continues the focus on *Maud*, reading this troubled and troubling poem as the climactic finale of Tennyson's experimentation with a uniformitarian poetics. *In Memoriam* speaker's attempts to shore up the self and forge a meaningful, forward pathway—a 'way of the soul'—out of a Lyellian non-progressive landscape bereft of meaning in the hope of regaining the narrative of self so cruelly interrupted by Hallam's death. *Maud*'s speaker, however, not only inhabits a Lyellian landscape, he internalises that landscape; it becomes the landscape of his psyche. Critics have long recognised the distinctiveness of *Maud*; its oddness is often felt but has proved difficult to characterise. This chapter approaches this oddness via Bakhtinian critical theory. It argues that the uniformitarian 'I' of the poem is one that is realised in the present and is therefore an unstable 'I' barely able to maintain monologic wholeness or narrative purpose. In *Maud*, Tennyson follows Lyell's uniformitarianism to its logical conclusion, taking its speaker to the limits of selfhood. The result is a poem that is not dialogic in itself but that performs a dialogic moment by positing the monologic consciousness of the speaker against a geologically and linguistically unstable world where meaning, like the geological agents that uniformitarianism brings into relief, is in constant process in the present. The chapter demonstrates how *Maud* pre-empts the appropriation of Lyell's uniformitarian methodology by linguists in the 1860s, an appropriation that profoundly disrupted conventional mid-nineteenth century perceptions of language as Adamic and teleological, and that led the way for the development of twentieth-century structuralism and post-structuralism. Such a development, the closing chapter argues, is prophetically figured by Tennyson in *Maud*'s remarkable uniformitarian poetic experimentation.

CHAPTER 2

Ida's Footprint in the Sand: *The Princess*, Geology and the Extinction of Feminism

We thought, when we began to read 'The Vestiges,' that we could trace therein the markings of a woman's foot.
Adam Sedgwick, "Vestiges of the Natural History of Creation."
Edinburgh Review, CLXV (1845): 2.

I went on Saturday last to a party at Mr. Murchison's house, assembled to behold tortoises in the act of walking upon dough. Prof. Buckland acted as master of the ceremonies. There were present many other geologists and savants [... and] on endeavouring to take them up it was found they had stuck so fast to the piecrust as only to be removed with half a pound of dough sticking to each foot. This being the case it was found necessary to employ a rolling pin, and to knead the paste afresh; nor did geological fingers distain the culinary offices. It was really a glorious scene to behold all the philosophers, flour besmeared, working away with tucked-up sleeves.[1]

John Murray's letter recounting the notoriously flamboyant William Buckland (along with other geologists) endeavouring to encourage tortoises to walk across a "piecrust" sets up a humorous mismatch of gender expectations. The purpose of Buckland's experiment was to try to establish what kind of creature could have made the fossil footprints discovered

---

[1] John Murray III to his father, Jan 23 (1828) *John Murray III, 1808–1892: A Brief Memoir*, John Murray IV (1919), 7–8.

© The Author(s) 2017
M. Geric, *Tennyson and Geology*, Palgrave Studies in Literature, Science and Medicine, DOI 10.1007/978-3-319-66110-0_2

in 1828 in the Corn Cockle Muir quarry near Dumfries in Scotland. The comedy of Murray's recollection comes from the conjunction of geologists, dough, rolling pins and flour—geologists engaged in the work of kitchen maids. The comic elements of Tennyson's poem, *The Princess* (1847), work similarly on gender expectations, yet they hinge not so much on the cross-dressing male heroes but rather, as the Canadian critic Samuel Edward Dawson enthusiastically attempted to show, from the incongruous conjunction of women and knowledge—the young men's "profound inward amusement at the weaknesses of the fair denizens of the female university".[2] What also connects *The Princess* and Buckland's experiment, however, are fossil footprints themselves and the provocative evidence of prehistoric life that such fossils indicate. For Ida, as the poem reveals, sees her feminist legacy as a "sandy footprint" which she hopes, like the fossil footprints from the distant past that fascinated contemporary geologists, will, in time, "harden into stone" (III, 254) to leave a lasting impression.

That *The Princess* is peculiarly of its time has been noted by critics ever since its publication. The poem's dressing "up poetically" (Conclusion, 6)—via Tennyson's "gorgeous lyric gift"—what may have seemed the universal naturalness of gender divisions in mid-nineteenth century culture, can now appear, as Eve Kosofsky Sedgwick suggests, "frothing-at-the-mouth mad".[3] Out of its era, the poem's gender ideologies stand out in stark relief like the bones of a particularly vicious—and possibly not entirely extinct—prehistoric 'monster'. Herbert Tucker reads the poem's 'medley' form as enabling the rarefied play between ideology and artifice: the "fabulous story line exposes ... the collusion of prevailing social codes with Tennyson's idyllic artifice, which by substituting linguistic structures for social ones masks the force whereby those codes prevail".[4] Readings of the poem's science, however, have tended to be less concerned with ideology and much more optimistic in terms of its representation of feminism. Virginia Zimmerman's geological reading, for example, suggests

---

[2] Samuel Edward Dawson, *A Study of Lord Tennyson's Poem, The Princess* (1882), 24. Tennyson, replying to Dawson's study of *The Princess*, demonstrates his agreement with Dawson's reading of the absurd mismatch of women and knowledge: "You have seen" he writes "amongst other things that if women were ever to play such freaks, the burlesque and the tragic might go hand in hand" (*Memoir*, I, 256).

[3] Eve Kosofsky Sedgwick, "Tennyson's Princess: One Bride for Seven Brothers" (1985) in *Tennyson*, Rebecca Stott ed. (1996), 182.

[4] Herbert F. Tucker, *Tennyson and the Doom of Romanticism* (1988), 356.

the importance of uniformitarianism for figuring a gradual movement towards gender equality: "Ida marries the prince but maintains her faith in uniformitarianism"; the Prince is Ida's "perfect mate" and between them "They present a pair of equals who will work together in small ways to affect change for women."[5] Equally optimistic, Rebecca Stott, examining its engagement with Chambers's *Vestiges of the Natural History of Creation* (1844), argues that the poem extends and valorises the mixed-gender conversations on science that *Vestiges* triggered: "*The Princess* is a *conversazione* as well as medley; it is a social event" and it "speaks for a radical politics that is both feminist and utopian". Stott concludes: "The conversation between the Prince and the Princess spills over the end of the poem's supposed closure."[6] This present reading also thinks about the importance of *Vestiges* for the poem, along with other key geological discourses, and particularly Hugh Miller's *The Old Red Sandstone* (1841). However, returning to a focus on ideology, it argues that the poem harnesses geology's material evidence (the fossil record) to assert a highly conservative view of the rightness and naturalness of gender inequalities. Far from facilitating a conversation beyond the poem, the poem's geology works to divisively close down the conversation, confining the Princess and her feminism firmly within the bounds of the fairy tale fantasy. Against geology's hard facts and visible evidence, Ida's claims for the rights of women are paper thin, merely "theories out of books" (Conclusion, 35)—just as Edmund Burke had earlier seen discourses on the 'natural' rights of man as "chaff and rags and paltry blurred shreds of paper".[7] As Hugh Miller suggested in terms of the pre-eminence of geological evidence, "It is well to return to the record, and to read in its unequivocal characters the lessons which it was intended to teach" as they offer "demonstrative evidence which the metaphysician cannot supply" (*ORS*, 102). *The Princess* works, via geology, to cast feminism as an aberration, a fad that has itself its origins in the female characters' misreadings of geology. Thus, far from endorsing feminism, *The Princess* engages with some of the most popular science writings of the day to silence it, to figure feminism as extinct and to consign it, once and for all, to the rubbish heap of history's most absurd and implausible creations.

---

[5] Virginia Zimmerman, *Excavating Victorians* (2008), 93.

[6] Rebecca Stott, "Tennyson's Drift: Evolution in *The Princess*" in *Darwin, Tennyson and Their Readers*, Valerie Purton ed. (2014), 32.

[7] Edmund Burke, *Reflection on the Revolution in France*, L.G. Mitchell ed. (1993), 86.

By the end of the embedded story Ida is sorely and suitably reduced; she is besieged by the force of an illogical but inescapable ideology which insists that her biology cancels out the logic of her arguments for women's higher education. In "hatred of her weakness" and "bent with shame" (VII, 15), she becomes "meek" and "mild" (VII, 210–1) admitting that "she had failed / In sweet humility; had failed in all" (VII, 213–4). As if this humiliation is not enough, she calls herself "Queen of farce! / When comes another such? never, I think, / Till the Sun drop dead" (VII, 228–30). Ida has made herself farcical in her insistence that women can be educated to the same degree as men, as to try do so, in the rhetoric of the poem, is to attempt to make women exactly what they can never be—men. As serious students, the women only offer a poor impersonation of men, equivalent to the comically dragged-up adventurers whose purpose is precisely to mirror the women in order to show them what they seem not to see: just how farcical their impersonation of men makes them look. The comedy may appear to centre on the men—"What! are the ladies of your land so tall?" (II, 26), "I sang, and maidenlike as far / As I could ape their treble" (IV, 73–4), "my voice / Rang false" (IV, 102–3)—but the joke is on the women, as Dawson made clear; "When turning from [her natural functions], she [woman] aims to play a part to which she has not been adapted, the moment her theories are put into practice she necessarily becomes absurd; and this, combined with the attractions of her sex, excites uncontrollably the sense of humour in man."[8] Ida's punishment for attempting to establish a university for women is a critical breakdown; "Her voice / Choked, and her forehead sank upon her hands, / And her great heart through all the faultful Past / Went sorrowing in a pause I dared not break" (VII, 231–3). She prays that the Prince will not judge the women's cause "from her / That wronged it" (VII, 220–1), and when the Prince consoles her by saying "Blame not thyself *too much*" (VII, 239, my emphasis), we get the distinct impression that she will not be allowed to forget her faults too quickly.[9] But Ida, of course, is not a real woman, she is a fairy-tale Princess, and, as her name teasingly hints, Ida is supposed to be ideology

---

[8] Dawson, *A Study* (1882), 22–3.

[9] As early as 1894, the critic Stopford A. Brooke recognised a certain "lordliness" in the Prince's "lecture on the woman and the man" at the end of the poem, which he suggested "belongs to Tennyson's attitude on the subject, and which makes me dread that Ida in after years lost a good deal of her individuality". See, Stopford A. Brooke, *Tennyson* (1898), 180.

personified. Ida gives shape to Lilia's half-formed feminist discontent and as an embodiment of feminist ideology, she stands in direct opposition to nature. It was "ill counsel" which "had misled the girl" (VII, 226); she had been "fed", we are told, "theories" (I, 128) by Lady Psyche and Lady Blanche, who, akin to the maiden aunt of the framing narrative, are themselves "crammed with theories out of books" (Conclusion, 35). In order to become a 'real woman', Ida must flesh-up and accept the dictates of her biology, as by ignoring her natural instincts she misses "what every woman counts her due, / Love, children, happiness" (III, 227). The choice is clear; Ida is either pure idea, in other words, non-existent and nothing biologically useful at all, or she is real, fleshy woman with no 'idea' at all.

This chapter reads *The Princess*'s geology as operating to shore up conservative ideologies about the importance of maintaining gender divisions in a time of both social discontent and unprecedented discovery in the earth sciences. The poem's gender message works through an internal logic based on contemporary geological notions of serial creation. What Tennyson finds in geology is a language and system with which to describe an incontrovertible natural hierarchy that repeats itself in every manifestation of creation. Geology serves to correct Ida's faulty feminist logic and to reinstate the natural order of both species and gender hierarchies. Ida's hope is to "lift the woman's fallen divinity / Upon an even pedestal with man" (III, 207–8), and she imagines her feminist legacy as geological remains; the "sandy footprint harden[ed] into stone" (III, 254).[10] In these aspirations, she seeks to set a precedent in the past for women's equality in the future. She seeks to leave a solid and tangible legacy which is meant to supply what is currently missing in the past and present—to set in stone, in other words, an exemplar of feminine heroics for future generations. Thus, *The Princess* forms its rhetoric around a reactionary summoning of the past. It is a poem that illustrates the seminal importance in the present of the continuity of power relations, not only in the figures of Sir Ralph, Sir Walter Vivien and his son Walter, but in each age of creation, which, despite the discontinuity suggested by serial creations, are all part of the 'one act of creation' that abides by an immutable and divine law.

---

[10] A poetic rhetoric of progress incorporating the idea of a "footprint on the sands of time" that may have influenced Tennyson can be found in Longfellow's "Psalm of Life" (1838).

GEOLOGY'S 'MASTER EXISTENCES'

The story of Princess Ida is told to "kill / Time" (Prologue, 200–1). It is the summer version of the game played in winter by the boys at university in which each boy narrates a different part of the story in turn. Lilia demands, "Kill him now / The tyrant! Kill him in the summer too" (Prologue, 201–2). Thus, the boys embark upon the construction of a story, an improvised medley, which they cast as a summer version of *The Winter's Tale*: "we should have him back / Who told the 'Winter's tale' to do it for us. / No matter: we will say whatever comes" (Prologue, 230–2). The tale that unfolds is a parable, albeit an "implausible parable".[11] The parable 'kills time' in its claim to be universal, as its message is ostensibly perennial. It is meant to impart a 'natural' wisdom, in this case, concerning the futility of trying to exceed what the text sees as the natural and biological limits of femininity. To assert its universality, like Shakespeare's *The Winter's Tale*, the temporal setting of the story is indeterminate. It is a fairy-tale medley which sets a chivalric and medieval past beside contemporary sensibilities and science (the geological expedition). The aim of the framed narrative is to kill time amid the rapid progress of the age, which is represented in the framing narrative's science and technology. And by killing time the parable reinforces the core truths that the poem works hard to suggest are as valid in the present as they have been in the past and will be in the future. The parable is meant to correct Lilia's aberrant feminist impulse and it works, as once it is told she becomes agreeably passive: "Lilia pleased me, for she took no part / In our dispute" (Conclusion, 29–30). In order for the idiosyncratic gender ideologies of the time to be passed off as natural, however, there needs to be a precedent for them in the past. Their universal rightness can only be convincingly asserted if they can be demonstrated to have always existed. Mobilising geology and its fossil evidence, the poem sets out to do just this.

The poem's speaker enters the parable via the present. He passes from the strange site of the Mechanics' Festival at Vivian Place—the frolicking "thousand heads", the "shrieks and laughter", the science, technology, song and dance, all "smacking of the time" (Prologue, 59, 89), to the ruined Abbey. With him he carries the chronicle that records the heraldic

---

[11] Tucker, *Tennyson* (1988), 353.

past and the history of Sir Ralph and the "warrior lady" who gave battle in defence of her besieged castle. The chronicle is an important link between the present and the past, but it also has another function. As "half-legend, half-historic" (Prologue, 30), it draws attention to its literariness and to the past as a medley of fact and fiction. If it is half-legend and half-history it can be confidently assumed that Sir Ralph, made real in the present by his statue in the Abbey grounds, represents the historic half, while the nameless "warrior lady" represents that which is legend. She is a "miracle of women", a "miracle of noble womanhood!"; her "stature" we are told, is "more than mortal", suggesting, presumably, that she is not a real woman at all. "So sang the gallant glorious chronicle" (Prologue, 35, 48, 40, 49) the speaker comments with an air of scepticism which suggests we read "gallant" as indicating flattery and embellishment designed to ennoble rather than to accurately record.[12] Thus, early on in the poem a tension is set up between cultural interpretation and material reality—a tension which is meant to indicate a subtle difference between art and nature and between literary versions of women and women themselves. This can be seen at the end of the poem when Ida capitulates and Walter remarks, "I wish she had not yielded!" (Conclusion, 5). Here, a wistful fondness for imagining women to be what they can never actually be (for dressing them up in a heroic garb that they can never in reality carry off) implies, in one sense, what Ida has to learn: that it is not men who thwart women's ambitions (as the male idolisation of women reveals their desire for women to be more than they are), but rather it is written into the biological nature of women that they must always fall short of expectation.

If the chronicle is part literary invention, a much stronger and more authentic link to the reality of the past is found in the material stature of Sir Ralph, which preserves and represents the past in its conflation of the

---

[12] The fifteen lines of "O Miracle of women" were added to the final edition of the poem published in 1851. Like the songs added to the third edition because "the public did not see the drift" (*Memoir*, I, 254), they must have seemed necessary. Edgar Finley Shannon, thought them to "set forth a noble prototype for his Princess [... and to] set the stage more thoroughly for Lilia's demand that the story be resolved by a fight and a subsequent clash of pseudo-medieval knights". See, Edgar Finley Shannon, Jr., *Tennyson and the Reviewers* (1967), 132. They may have also helped, as I argue here, to emphasise the difference between ideology and nature, between, in other words, what the poem sees as culture's fiction of women and the reality of women.

physical characteristics of a specific individual (the likeness of Sir Ralph) and geological substance (the stone itself). Where the warrior lady is "miracle", Sir Ralph's statue is moulded on that which was once real and animate, the flesh and blood hero of the chivalric past. This solid, material authenticity, which the chronicle lacks, places the statue in relation to the ossified bones of those previous great 'masters of the world' that Ida and the Prince encounter on the geological expedition. Much like the statue, geology reveals a past in which the previous masters of the earth are memorialised and their bones transformed by geological processes to remain in the present as monumental and irrefutable evidence of the masters of a previous age. The sheer material substance of fossil evidence—the "bones of some vast bulk that lived and roared / Before man was" (III, 277–8)—testifies to a natural materiality that the chronicle's embellishments cannot match. The bones of the vast bulk also draw attention to contemporary notions of serial creations, as gaps in the fossil record appeared to suggest that there had been successive creations of life on earth followed by mass extinctions—a theory adhered to by Cuvier. Hugh Miller's geology seems to be particularly pertinent here, and there are strong linguistic links between *The Princess* and Miller's *The Old Red Sandstone* (1841).[13] Miller writes in the ubiquitous language of hierarchy common to early and mid-nineteenth century geology. He sees each geological age as having its own "master existence" which makes itself distinct in its fossil remains (*ORS*, 44, 74, 224, 241). Such characterisation was not uncommon. Robert Chambers's *Vestiges*, even while adhering to a developmental hypothesis, saw each age as characterised by its "master-form or type".[14] Miller's prehistory allowed for progressive serial creation by suggesting that "races higher in the scale of instinct" succeed one another. Thus, when contemplating the evidence of geology, the "bones" of the "vast bulk that lived and roared / Before man was" (III, 77–8), Ida is paraphrasing Miller's assertion that "succeeding creations of the earth [existed], ere man was" (*ORS*, 274, 102). And a similar image is given by the Prince, who, arguing against his father who urges the use of force to conquer Ida, conjures up the extinct past:

[13] For Tennyson's reading of Miller's *Old Red Sandstone* (1841), see Chap. 1.

[14] Robert Chamber, *Vestiges of the Natural History of Creation*, 1844 (1994), 84.

I would the old God of war himself were dead,
Forgotten, rusting on his iron hills,
Rotting on some wild shore with ribs of wreck,
Or like an old-world mammoth bulked in ice,
Not to be molten out. (V, 139–43)

Here the "God of war" that rules the present is analogically linked with the now extinct mammoth who was the 'master existence' of a previous age.

The mammoth preserved in ice had been a startling discovery noted in a number of geological sources. *Vestiges* reported that "a specimen, in all respects entire, was found in 1801, preserved in ice, in Siberia".[15] Lyell too notes the elephant found "in a mass of ice on the shore of the north sea" (*PG*, I, 54). But Miller, more poetically, writes of "The single elephant, preserved in an iceberg beside the Arctic Ocean, [which] illustrated the peculiarities of the numerous extinct family to which it belonged, whose bones and huge tusks whiten the wastes of Siberia" (*ORS*, 157). Miller also makes an explicit link between extinct elephants and previous masters of the world. At length, he writes, "after races higher in the scale of instinct had taken precedence in succession, the one of the other, the sagacious elephant appeared, as lord of that latest creation which immediately preceded our own" (274). Equally, the poem encodes Miller's view of each successive creation exhibiting its own hierarchy in which the 'lord' or 'master existence' of the respective age (in this case the mammoth) is crudely estimated by the "vast bulk" of its remains. If such is the measure of the master existence, then King Gamma's twin sons, "those two bulks at Arac's side", and Arac himself, "lord of the ringing lists" who rains down blows "as from a giant's flail" (V, 488–91), represent the master existence of an Anthropocene age where brute force has ultimate sway. Even Ida concedes to the pre-eminence of brute force, for it is the victors in battle who the women will revere; they will be "The sole men we shall prize in the after-time", whose "very armour hallowed" and whose "statues / Reared" are "sung to" (V, 402–4).

Bound up with the imagery of fossil remains is Sir Ralph's statue, which speaks of the master existence of the present age and of a natural

---

[15] Ibid., 130.

superiority figured through physical presence and strength. Unlike the chronicle, the statue (like fossil remains) offers irrefutable evidence, ostensibly outside cultural construction, of the natural inequality of gender. In lieu of bones as yet unfossilised, the 'lord' of the present creation, 'man', is fittingly represented in Sir Ralph's stone effigy. Toppled and crumbling though it may be, it represents the continuity of the 'master-form' in the present era—a continuity that flows through Sir Walter Vivian and his son Walter. The statue of Sir Ralph, like the evidence of geology itself, is meant to be a powerful and irrefutable indicator of the natural, historical order. It is a silent censor, cancelling gender equality with a material authority that goes before and beyond the authority of the chronicle. Such material, non-linguistic evidence makes a nonsense of the reams of paper stamped with theories on the rights of women on which Ida has been "fed". Fittingly, where Sir Ralph is named and memorialised, there is no statue for the nameless warrior lady, no evidence of her real existence beyond the textual space. As if in tacit recognition of this, Lilia's robing of Sir Ralph's statue in women's clothing can be seen as an attempt to redress a past in which women do not tangibly figure. However, the clothes merely reaffirm the pitifully slight impression women make on the hard, material world. They are the trivial accessories that, in the text's logic, characterise women themselves. The text begs the question: what remains have women left in the past? As Cyril points out in a similar inquiry, "when did woman ever yet invent?" (II, 369). Psyche's lecture plucks out a range of notable women mainly from the history of literature and legend. Yet any material evidence of women's equality with men is starkly lacking, suggesting that women's place in history remains consistent; it is nominal and homogenous. Women need not be commemorated individually; instead, statuary representations of them can be confined to either the celebration of woman's generic form, or to allegorical representation, and in both, it is generally the artist who is commemorated and not the individual woman.

Just such allegorical representation is found in the "woman-statue ... with wings" (I, 207) that the Prince and his companions confront as they approach the all-female university. The statue's generic status exemplifies the representation of women as homogeneous. The winged "woman-statue" recalls the winged sphinx and the mythologies of women's subservience to their biology. In the poem's rhetoric, women are "truer to the law within", less reasoning and therefore less just: "Severer in the logic of a life" (V, 181–2). The winged statue associates women with the monstrous mythological amalgamations of women and birds, as Florian says, "who could think / The softer Adams of your Academe, / O sister, Sirens though they be, were such / As chanted on the blanching bones of men?" (II, 279–82). Ida is herself, of course, like the warrior lady of the chronicle, not real but a caricature; she is Lilia made "some great Princess, six feet high / Grand, epic, homicidal" (Prologue, 218–9). She is Lilia's childish waywardness magnified in order to demonstrate what her monstrous feminist reasoning might look like if realised. Where Lilia is the "petty Ogress" (Prologue, 156), Ida—putting that feminism into practice—is the miscreant sphinx/siren who cannot be reasoned with; "that iron will, / That axelike edge unturnable" (II, 285–6). And fittingly, the men fail to read the "inscription that ran along the front" of the "woman-statue" because it is "deep in shadow" (I, 207–10); it is inscribed, presumably, in a biological vernacular, too far below the pitch of reason. Consequently, the men enter into the dangerous, cloying realm of instinct, where sex and death provide the only rationale, albeit that the men manage to solve the riddle, destroy the university, domesticate the "sirens" and secure the reproduction of the species as each claims his own bride. Where the statue of Sir Ralph represents the individual man, the "woman-statue ... with wings" suffices in its representation of women who, it seems, never quite surface from the muted fluidity of their somatic world into the clear air of self-determination. The two statues point to an important difference between men's and women's relationship to history. Not only this, the 'woman' statue also forges a tie between women and birds that becomes important in the reading of the poem's figuration of geological remains.

KILLING TIME

There is a curious conflation of perceptions of time in *The Princess* that seems to be a part of the poem's 'medley' status. On one hand, the framing narrative sets in place the contemporary and the linear; this is Victorian England with its past represented by the Abbey ruins, and its future represented by the poem's science and technology. On the other hand, the framed parable, as already suggested, has no temporal anchor and is designed to impart timeless truths. Thus, the poem is a self-referential medley of incongruous parts, not just of genre (tragedy and comedy, realism and fantasy, narrative and lyric forms) and of people, objects and ideas, but also a medley of temporal references. The poem brings into close proximity many supposedly incongruous elements: steam engines and Gothic ruins, electric shocks and medieval knights, mechanics and landed gentry, geological remains and cultural artefacts and women and higher education.[16] What unites all these disparate elements, however, is the poem's geology represented most powerfully through Ida's perception of geological time.

Ida's contemplation of the remains of the "vast bulk" leads her to hope for progress: "As these rude bones to us, are we to her / That will be" (III, 279–80). What she fails to recognise, however, is that the fossil remains she chooses for her example speak not of continuous progress but rather of serial creations and extinctions. If, as Miller believed, serial creation was progressive, as "races higher in the scale of instinct" succeed one another, the future creation to which Ida looks for progress can have no relation, biological or intellectual, to the present, as nothing can be handed down through extinction (ORS, 102). Staunchly anti-evolutionary, Miller's adherence to serial creation helped him avoid developmental and materialist theories. Arguing against Benoît de Maillet's and Jean-Baptiste Lamarck's hypotheses of the transmutation of species, Miller maintained that fossil evidence suggested the existence of complex organisms even in the earliest known strata—thus demonstrating that any import of continuous progression was merely an illusion. Geology "furnishes no genealogical link to show that the existences of one race derive their lineage from the existences of another". Ida's problem is that serial creation suggests that if there is to be an improved womankind in the future, she could not have inherited her improvements from Ida, which makes Ida's exertion in the present profoundly futile.[17]

The Prince responds to Ida with a theological question pertaining to serial creation; "Dare we dream of that ... / Which wrought us, as the

---

[16]Virginia Zimmerman describes the poem, for example, as a "temporal medley, as well as a literary one". See *Excavating Victorians* (2008), 71.

[17]By evoking serial creation, Ida raises a similar problem to the one John Killham notes in *In Memoriam*. In Tennyson's elegy, 'man' is "'herald of a higher race', provided only that he followed the example time had brought to light concerning the rest of the animal kingdom". As Killham argues, how this progress might come about if each creation is "totally destroyed and re-created is difficult to see". See, John Killham, *Tennyson and The Princess: Reflections of an Age* (1958), 246.

workman and his work, / That practice betters?" (III, 280–2).[18] Ida's answer evokes the numinous and transcendental:

> To your question now,
> Which touches on the workman and his work.
> Let there be light and there was light: 'tis so:
> For was, and is, and will be, are but is;
> And all creation is one act at once,
> The birth of light: but we that are not all,
> As parts, can see but parts, now this, now that,
> And live, perforce, from thought to thought, and make
> One act a phantom of succession: thus
> Our weakness somehow shapes the shadow, Time;
> But in the shadow will we work, and mould
> The woman to the fuller day. (III, 303–14)

Here she effectively 'kills time', envisaging "one act" of creation in which all "is" already foretold in the divine mind—a perception that is intrinsically teleological and distinctly Millerian in its phraseology. Miller also saw the whole of creation as 'one act' and argued that while the fossil record and the idea of serial creations and extinctions appeared to suggest discontinuity, the disparate parts of creation made up a cohesive, divinely organised whole. He asks his readers to recall Milton's "body of Truth ... hewn in pieces, and her limbs scattered over distant regions, and how her friends and disciples have to go wandering all over the world in quest of them". Similarly, the geologist, exercising reason and imagination, must unite the parts of creation that make up the whole:

> There is surely something very wonderful in the fact, that in uniting the links of the chain of creation into an unbroken whole, we have in like manner to seek for them all along the scale of the geologist;—some we discover among the tribes first annihilated—some among the tribes that perished at a later period—some among the existences of the passing time. We find the present incomplete without the past—the recent without the extinct. There are marvellous analogies which pervade the scheme of Providence, and unite, as it were, its lower with its higher parts. (*ORS*, 45)

---

[18] James Eli Adams reads these lines as responding to Ida's hope for evolutionary progress. See "Woman Red in Tooth and Claw: Nature and the Feminine in Tennyson and Darwin," in *Tennyson*, Rebecca Stott ed. (1996), 99.

Every part of creation has its place in the finely tuned order of divine creation. That order appears successive, but despite this, it is not resultant on what has gone before. New creations might appear to be improvements, yet such improvement does not entail a continuous progress in living species. Rather, "The perfection of the works of Deity is a perfection entire in its components, and yet these are not contemporaneous, but successive: it is a perfection which includes the dead as well as the living, and bears relation in its completeness, not to time, but to eternity" (*ORS*, 45).

Ida's 'one act' of creation, which for those within it ("we that are not all") appears as in parts, echoes Miller's use of metaphor. The passing geological ages Miller sees as "scenes" that shift as might the scenes of a theatrical production but with each new scene introducing new characters. And significantly, to kill time (what Ida calls the "phantom" "shadow" through which the individual must live) Miller evokes Shakespeare's *The Winter's Tale* as a way of representing the individual's experience of linear time as encompassed by geological timelessness.

> The scene shifts, as we pass from formation to formation; we are introduced in each to a new *dramatis personæ*; and there exist no such proofs of their being at once different and yet the same, as those produced in the *Winter's Tale* to show that the grown shepherdess of the one scene is identical with the exposed infant of the scene that went before. Nay, the reverse is well nigh as strikingly the case, as if the grown shepherdess had been introduced into the earlier scenes of the drama, and the child into its concluding scenes. (*ORS*, 40–1)

If the fossil record was directional and progressive, "we would necessarily expect to find lower orders of fish passing into higher, and taking precedence of the higher in their appearance in point of time, just as in the *Winter's Tale* we see the infant preceding the adult" (*ORS*, 44). According to Miller, no such expectation is met with in the fossil record. In fact, he argues, as Lyell had done before him, that complexity abounds in all strata, from the newest to the oldest; both find—in other words—that the adult shepherdess often comes before the infant. Human time, however, must unfold as do the scenes of *The Winter's Tale*; the infant must precede the shepherdess or else the human narrative is lost. For those within the divine plan, its narrative appears in process. However, it does not 'progress', as creation is already complete in every part at all moments within the divine mind.

Miller kills time and its progressive import by conjuring Shakespeare's *The Winter's Tale*. Ida also kills time, and the fairy tale in which she herself is one of the major "*dramatis personæ*" is designed not only to "kill" the "tyrant time" but, as the Prologue announces, to be a summer version of the "Winter's tale" (Prologue, 231). If the framing story offers the linear narrative of the human perception of time as Shakespeare's play does (the real world in which Lilia exists), then the framed story of Ida offers Miller's timeless view of geological creation as a story already told, and one in which everything is already in existence and is part of a single act. And Tennyson hints at this when he explains that "there is scarcely anything in the story which is not prophetically glanced at in the prologue" (*Memoir*, I, 251). What the characters cannot see because they are part of the whole, the reader, as the omnipotent observer of the whole, sees or is meant to sense. *The Princess*, as a medley, makes it the job of the poet/narrator and reader to bind together the "scattered scheme" (Conclusion, 8), just as the geologist must bring the temporally disparate parts of natural creation together to understand the whole. A similar reconstruction occurs in comparative anatomy, as already suggested in Chap. 1, where the geologist and palaeontologist Gideon Mantell speaks of the "medley of bones" that must be constructed into a whole.[19] *The Princess*'s medley of parts, makes sense of the past and present—"every clime and age / Jumbled together" (Prologue, 16–7)— and reveals, in Miller's words, the "marvellous analogies which pervade the scheme ... and unite, as it were, its lower with its higher parts". The poem's Millerian cohesion is given in its dual perception: the timelessness of the parable, and the linear progress of the framing narrative, which together offer a reverent and geologically oriented awareness of the unchanging truths of a system in process while it is yet complete. The poem's temporal play is complex; this is, after all, a poem in which the speaker informs the reader at the very conclusion that the story they have just read/heard is now to be dressed "up poetically" (6). Thus—in what seems to be absurdly clever facet of *The Princess*'s manipulation of readers' perceptions—the end is reached only for the reader to discover that the poem has yet to be written; the story, it seems, is still in process, even while it is complete.

---

[19] Mantell, *The Wonders of Geology Or, A Familiar Exposition of Geological Phenomena* (1838), I, 127.

Miller's dual perception foregrounds how the unity of nature is not found in the progressive continuity of species, which for Miller is an illusion. Rather it is in the natural order that is consistently visible in all the separate 'scenes' of creation, and that can be apprehended when all creation, successive as it appears, is viewed as complete. Each scene, while complete in itself, is also connected to all others in the unfathomable mind of the "adorable Architect" (*ORS*, 96). Each scene leaves in its remains the marks of the "master existence", and within each scene relationships to the master existence are pre-ordained, fixed and immutable. Miller explains how within any given species there is a certain amount of room for movement. In the human species, for example, "we find full-grown men of five feet, five feet six inches, six feet, and six feet and a half", but this does not mean that the "race is rising in stature, and that at some future period the average height of the human family will be somewhat between ten and eleven feet". Such variance is merely gradation, and "gradation is not progress" (*ORS*, 43). Thus, while Miller suggests that there is progression in "the scale of instinct" of successive creations, there can be no progress out of the bounds of species, only movement up and down a scale within the limits of species boundaries.

For the human species, however, this scale has a moral as well as a physical dimension, as it can describe the varying states of being between the most 'primitive' and the most 'perfect' within a species. Psyche's lecture makes a distinction between past master existences and the masters of the present creation in terms of a sequence: "then the monster, then the man"; but she then describes the moral condition of man as existing at any moment in all its forms.

> Tattooed or woaded, winter-clad in skins,
> Raw from the prime, and crushing down his mate;
> As yet we find in barbarous isles, and here
> Among the lowest (II, 101–5)

There is no linear progress towards civilisation here, only gradation, so that any moral condition can be found in any era of mankind. Thus, there is the possibility of moral perfection in the present but no possibility of evolution beyond that fullest potential.[20] The moral perfection of

---

[20]This is a Lyellian view of human potential that he adhered to in order to evade the import of evolutionary theories, as suggested in Chap. 3.

the individual is achievable in the present because the potential for moral perfection is already immanent in the human species. In the gender divisions of the poem, gradation helps to create the notion that men and women are not "equal, nor unequal" (VII, 285)—they are not different from each other—rather together man and woman make up a single example of the species. As the Prince asserts, the perfection of the human species is realised in "The single pure and perfect animal / The two-celled heart beating, with one full stroke, / Life." (VII, 288–10). Women achieve their perfection when they are subsumed into the master existence outside of which they are incomplete and cannot represent the species. As the Prince makes clear, "woman is not undevelopt man" (VII, 259); thus, she can never hope to develop herself to become equal with man. Rather she is properly part of man; woman must "set herself to man, / Like perfect music unto noble words" (VII, 269–70). She is the accompaniment to a narrative devised by him, as Ida is the creation of the male narrative pieced together to make a whole from the seven narrated parts. Women are part of the story of the species, but they are not representative of it. They must sing to the statues of men, and within those statues, women find their apotheosis. But women can never hope to exemplify the master existence of which they are but a part. In the anti-progressive view of the fossil record in which hierarchal relations are set in place by immutable laws, the age of the human species speaks of the age of men, and only men, who leave the mark of their supremacy in their statuary remains, like the fossil bones of previous master existences.

Ida's reverent vision of geologic time is what marks her as suitable for rehabilitation, but her ambition to leave a solid legacy in the "footprint harden[ed] into stone" is unrealistic, just as her stature and nobility would be outside the timeless sphere of the parable. In the world outside the parable, she is Lilia, the real woman, who exists in the linear timeframe of the enclosing narrative—a world in which the child precedes the adult, the infant the shepherdess. Ida must step out of the parable into the stream of time and reduced from her legendary status, she must relinquish her grandiose desire to set her legacy in stone and join instead the "soft and milky rabble of womankind" (IV, 290).[21] She must "melt" (VI, 268) into the undifferentiated fluidity assigned by wifedom and motherhood, and resign herself to the domestic heroism allotted to mortal woman. Neither Ida nor Lilia, in the end, are able to kill time. Like

---

[21] Robert Bernard Martin, *Tennyson: The Unquiet Heart* (1980), 314.

Hermione in *The Winter's Tale*, who is metaphorically turned to stone once banished from the filial relations that animate her, time continues as the statue that represents her physically ages. Hermione is released from her incarceration in stone once she is accepted back as the king's wife and re-assimilated into the generational progress that the play suggests for Miller. Similarly, Ida, whose isolation and stone-like coldness are self-imposed, must hear the shepherd's song and "Come down" from her "mountain height" (VII, 177–8) to take up her reproductive role. As a woman, her legacy must be a biological one; not one carved in stone, as Sir Ralph's is, nor one signified by the hard remains that memorialise past master existences, but the soft and fleshy legacy of children that women leave behind. As Hermione responds to Paulina's address, "'Tis time. Descend, Be stone no more" (5. 3. 11) and steps down from the pedestal to become animate again, so Ida exchanges her claim to monumental remembrance in the "even pedestal with man" (III, 208), for the valleys of love—"all her labour" (the fashioning of her great deeds into a solid memorial, abandoned) "a block / Left in the quarry" (VII, 215).

IDA AND PSYCHE

Many critics have pointed to the significance of Tennyson's reading of Chambers's *Vestiges* (1844) for *The Princess*.[22] *Vestiges*, written three years after the publication of Miller's *Old Red Sandstone*, offered what Lyell and Miller must have deplored: a popular, accessible, homegrown grand narrative of the history of the earth and its species based on the assumption of evolutionary development—theories linked to materialism, atheism, and by implication, social discord and moral decline. The evolutionary sections of *The Princess*, as John Killham has argued, were probably draw from Lyell's *Principles* rather than *Vestiges*, as "the facts underlying those stanzas which are most justly called evolutionary can be found in Lyell's work".[23] However, references to the nebular hypothesis in *The Princess* suggest the influence of *Vestiges* as Chambers based his broad history of creation on the premise of the theory, and although Lyell also discusses it, it is Chambers's particular use of the hypothesis

---

[22] See, specifically, Milton Millhauser, "Tennyson's *Princess* and *Vestiges*" (1954). John Killham, *Tennyson and The Princess* (1958), 252–66. Rebecca Stott remarks that "*The Princess*, though conceived around 1839, was largely written whilst the controversy over *Vestiges* was at its peak, forged out of mixed-sex conversations that dominated the talk of dinner tables in 1845 and 1846." See, "Tennyson's Drift," in Purton ed., *Darwin, Tennyson* (2014), 23.

[23] Killham, *Tennyson and The Princess* (1958), 252.

to bolster his materialist theories that feature in the poem. Popularised by Laplace and Herschel, the nebular hypothesis suggested ongoing, linear development with the Solar System being formed initially from condensed gaseous material, and, as Killham points out, it was often adopted by early evolutionists.[24] Tennyson knew of the theory before he had read *Vestiges*, as cancelled stanzas of *The Palace of Art* (1833) suggest.[25] However, for *The Princess* it was the controversial, materialist and evolutionary associations of the theory as offered in *Vestiges* that Tennyson draws attention to, and significantly it is Ida and Psyche who allude to the nebular hypothesis. As Rebecca Stott argues, the controversy surrounding *Vestiges* has a particular relevance for *The Princess*.[26] *Vestiges*, as James Secord discusses, was "more controversial than any other philosophical or scientific work of its time".[27] Its widespread popularity and the sensation it aroused made it known and accessible to women; it helped to make "science a shared interest" between men and women and it was "often given by men to women".[28] It fuelled social conversation across the gender divide and invited women to "talk familiarly" on the most speculative of subjects, offering them a chance to ruminate over the broadest questions of earth's history, the origin of creation and the impact of geological discoveries for theological debates.[29]

---

[24] Ibid., 234.

[25] Ibid., 233–4.

[26] Rebecca Stott, "Tennyson's Drift: Evolution in *The Princess*" in *Darwin, Tennyson and Their Readers*, Valerie Purton ed. (2014).

[27] James Secord, *Victorian Sensation* (2003), 1.

[28] Ibid., 169.

[29] Ibid., 166. Secord relays a telling dinner conversation between Sir John Cam Hobhouse and Ada Lovelace. Sir John, who suspected Lovelace to be a "witch", reported a conversation "that involved issues relevant to the final chapter of *Vestiges*. Lovelace questioned the validity of 'the common argument in favour of a future state' derived from the intimations of immortality. However, 'she would not say so much to anyone, except very privately,' lest she deprive them of 'the consolation of belief'. Her own thoughts were 'only doubts'—to be dogmatic would be 'presumptuous'. Even so, Hobhouse was taken back, for these were 'subjects which few men, and scarcely any women, venture to touch upon'". A link between *Vestiges'* materialism and outright atheism is here made clear, and Secord points out that "Lovelace's behaviour" in discussing such subjects "was simultaneously identified as gendered, aristocratic, and eccentric. Certainly any gentleman of science would have recognized that reference to such matters would have been out of place" (416).

Such questions, which had an implicit bearing on issues concerning the assumed hierarchies of class and gender, had been seen as unfit for discussion outside select male circles of the scientific and theological elite. *Vestiges* was guilty of disseminating ideas that had no place in publications easily accessible to women, and by inviting women into a conversation about the origins of humanity, *Vestiges* sinned not only against contemporary taste and decorum but also against femininity itself. By placing the nebular hypothesis in the mouths of the women educators in the poem, Tennyson was specifically associating them with *Vestiges* and its propagation of some of the most controversial developmental theories of the day. However, there is an important difference between Psyche's and Ida's approaches to the nebular hypothesis.

Psyche begins her lecture by describing the beginning of the universe (not only the Solar System) as rotating clouds of nebular material that condense into matter:

> This world was once a fluid haze of light,
> Till toward the centre set the starry tides,
> And eddied into suns, that wheeling cast
> The planets. (II, 101–4)

This is the starting point for a developmental system that extends to animal species and human culture and finally to the history of women, albeit that Psyche appears to lump together serial creation and gradation, as well as evolutionary theories. Later, addressing the Prince privately, Ida also refers to the nebular hypothesis: "There sinks the nebulous star we call the Sun, / If that hypothesis of theirs be sound" (IV, 1–2). The difference is significant. Where Psyche makes the hypothesis the foundation of her lecture—that from which all else follows—Ida shows signs of scepticism. Where Psyche asserts; "This world was once …", Ida takes the hypothesis for the *hypothesis* that it is. She resists the language of developmental progress and avoids the implications of materialism that they suggest. In this we are meant to see Ida as innately discerning, as able to exercise a womanly reserve that prevents her from disregarding the authority of the past for the first fashionable new hypothesis that comes along. Ida's more reverent approach to science is meant also to show the inbred superiority of her nobility—a notion endorsed by the poem's investment in the continuity of a benevolent aristocracy. Ida's scepticism matches her ability to kill time and see "all creation" as "one act", which also lends her science a sense of mystical reverence. Thus, she supplies the best model of womanliness in the poem. Ida, after all, was only subverted from her high path in the vulnerable years of her youth by the "Two widows, Lady Psyche, Lady Blanche" (I, 127). Her natural reverence, her perception of divine creation and its relation to human time, her anti-materialism and her uncertainty concerning the nebular hypothesis, are what save her from her "false" self and its faulty feminism.

A good gauge of the reception of *Vestiges* can be found in Adam Sedgwick's often quoted review. Sedgwick's tone reveals the anxiety that the book initiated in conservative circles about the import of materialist and developmental theories for both the social order and the gender hierarchy:

> If our glorious maidens and matrons may not soil their fingers with the dirty knife of the anatomist, neither may they poison the springs of joyous thought and modest feeling, by listening to the seductions of this author; who comes before them with a bright, polished, and many-coloured surface, and the serpent coils of a false philosophy, asks them again to stretch out their hands and pluck forbidden fruit—to talk familiarly with him of things that cannot be so much as named without raising a blush upon a modest cheek;—who tells them—that their Bible is a fable when it teaches them that they were made in the image of God—that they are the children of apes and the breeders of monsters—that he has *annulled all distinction between physical and moral*—and that all the phenomena of the universe, dead and living, are to be put before the mind in a new jargon, and as the progression and development of a rank, unbending, and degrading materialism.[30]

Psyche's knowledge of the nebular hypothesis shows her to be both up-to-the-minute in her reading and dangerously familiar with theories which were in themselves the prelude to the ideas of progressive development that *Vestiges* expounded. *Vestiges*, Sedgwick suggests, was dangerous because it was designed to attract the weak-minded and credulous, those for whom surface is all and depth is naught; and Psyche, unlike Ida, is the contemporary woman who might be impressed by the "bright, polished, and many-coloured surface" of *Vestiges*, while unaware

---

[30] Adam Sedgwick, "Vestiges of the Natural History of Creation", *Edinburgh Review*, CLXV (1845), 2.

of the "serpent coils of a false philosophy". Moreover, Psyche's connection with *Vestiges* did not go unnoticed by early reviewers of the poem. Samuel Edward Dawson had no trouble in recognising Psyche as a common type of *Vestiges* reading female: "We see her as the young mother, full of love for her babe, and of attachment to the Princess, taking up the nebula hypothesis in the same way as young women now, with only one child, three castles and too much leisure, take up willow pattern china, and ugly furniture, and dignify such pursuits with the name of culture."[31] The assumption was clear; the innate superficiality of women, when untempered by natural decorum and moral reserve, leads them into any absurdity. And, as Sedgwick implies, such superficiality precludes women from any serious study of science.

Psyche's lecture also alludes to the more dangerous expounding of *Vestiges* exemplified in the lectures of Emma Martin. John Killham's seminal examination of *The Princess* and its engagement with the Owenites and their feminism does not mention Martin. However, she was a notorious atheist and feminist speaker on the Owenite lecture circuits in the 1840s. She was renowned for her rhetorical skill and her reasoning ability and best known for her lectures and pamphlets renouncing Christianity and the Bible.[32] Significantly, in 1846, Martin lectured on *Vestiges*, although the content of the lectures is not known.[33] Her lectures attracted thousands and there was often fierce opposition to her exercising her right to free speech.[34] Her impact was such that Miller thought it necessary to ridicule Martin in the pages of his radical Evangelical newspaper, *The Witness*, in 1845.[35] Thus, the female lecturer, speaking on materialism and feminism, was a recognisable figure in the 1840s. For her association with Martin, Psyche suffers inordinately in the poem, her thorough humiliation being a measure of the severity of her transgression. She begins the story as an icon of female learning; her "on whom / The secular emancipation turns / Of half this world" (II, 268–70). However, she is quickly conquered by the men's remembrances of domestic scenes, which are meant to foreground the inevitable "clash"

---

[31] Dawson, *A Study* (1882), 43.

[32] For more on Martin see Barbara Taylor, *Eve and the New Jerusalem* (1983), 149–56.

[33] Secord (2003), 317.

[34] See, for example, "Mrs. Martin." *The Movement*, Jacob Holyoake ed., 45. (1845), 390.

[35] See Secord (2003), 318–9.

between "love and duty" (II, 273) that—in the rhetoric of the poem—will always compromise women and mark them as unsuitable for the professions or for public office. Psyche must be re-assimilated into the nexus of filial love and (removed from the lectern) she is reduced to a "Pitiful sight ... Like some sweet sculpture draped from head to foot, / And pushed by rude hands from its pedestal" (V, 53–5). If Ida has sought to "To lift the woman's fallen divinity / Upon an even pedestal with man" (III, 208), then Psyche is routinely cast from this pedestal in a re-affirmation of the text's message that a lasting representation of the individual 'woman of knowledge' is an anomaly in itself. Any sympathy elicited from Psyche's harsh treatment from "rude hands" belies the text's adherence to those immutable natural laws that insist that true power and knowledge is inscribed in solid remains, such as those of Sir Ralph or the "vast bulk" of more distant "master existences". This is the hard, implacable law that Psyche must learn.

Unlike Psyche, Ida displays her natural feminine sensibilities in her distaste for anatomical dissection and her reverence for the "holy secrets of the microcosm". In keeping with the dictate of the day that ruled that "Ladies could see skeletons and stuffed animals, but dissecting rooms were male preserves", Ida affirms that in the women's university, while there are schools for all subjects, dissection and vivisection are not permitted.[36] When the Prince remarks: "Methinks I have not found among them all / One anatomic", Ida replies:

> Nay, we thought of that,
> ... but it pleased us not: in truth
> We shudder but to dream our maids should ape
> Those monstrous males that carve the living hound,
> And cram him with the fragments of the grave,
> Or in the dark dissolving human heart,
> And holy secrets of this microcosm,
> Dabbling a shameless hand with shameful jest,
> Encarnalize their spirits: yet we know
> Knowledge is knowledge, and this matter hangs. (III, 290–9)

---

[36] Ibid., 251.

Ida is true to Sedgwick's assumption; she will not "soil her fingers with the dirty knife of the anatomist". Miriam Bailin suggests, "The possibility that women might be so engaged is raised in order to suppress it as unnatural—a violation of female nature—which fundamentally threatens the integrity of the male self and ultimately of the social order."[37] For Sedgwick, developmental theories were equally 'dirty', leading women to an immodest and unbecoming biological knowingness. Ida also demonstrates her reverential sensibilities when she desists from examining the "holy secrets" of the heart's "microcosm", exhibiting her innate constitutional unfitness, as Sedgwick suggests, for the perusal of such sacred evidence. She distains such knowledge even while she affirms, "yet we know / Knowledge is knowledge"; thus, she demonstrates her tacit understanding of the 'natural' limits of women's capacity for knowledge. Ida's forbearance plays into the ideology that sees in women a heightened moral sensibility, while it simultaneously illustrates an innate weakness that demonstrates their inherent inability to match men in the knowledge stakes.

The "holy secrets" of nature are Miller's concern too. To 'dabble' a "shameless hand with shameful jest" in the invisible microcosm of anatomy is to meddle irreverently with divine handiwork. Miller discussed the intricate patterned beauty of fossil fish: "nor does it lessen their wonder" he writes, "that their nicer ornaments should yield their beauty only to the microscope". Miller saw in the composition of fossil fish the signature of divine creativity which runs through nature. The "unassisted eye", he writes, "fails to discover the finer evidences of this unity: it would seem as if the adorable Architect had wrought it out in secret with reference to the Divine idea alone". The microscope, paradoxically, is a mechanical tool which opens up a spectacle of sacred import and, therefore, it must be used with an appropriate reverence for the divine splendour it discloses. Miller accentuates the awesome implications of discovering such stunning creativity with an analogy: "The artist who sculpted the cherry-stone consigned it to a cabinet, and placed a microscope beside it; the microscopic beauty of these ancient fish was consigned to the twilight depths of a primeval ocean" (ORS, 96).[38]

---

[37] Miriam Bailin, *The Sick Room in Victorian Fiction* (1994), 44.

[38] For interesting material on Tennyson's experience of the microscope through the geologist Charles William Peach, see, Lyall I. Anderson and Michael A. Taylor (2015) "Tennyson and the Geologists Part 1: The Early Years and Charles Peach" (2015): 348–50.

Remarkably, evidence of divine design has lain hidden and inaccessible in the incomprehensible depths of time long before human consciousness existed to fathom its significance, and even now it is rendered "secret" and remote from human comprehension by its minuteness. The geologist and the anatomist are here akin to the priest, as the secret store of knowledge hidden in the finely wrought artistry of prehistoric fish and in the "holy secrets" of the heart's microcosm show where the marks of nature's divinity and the scientific apprehension of it meet—a realm of knowledge naturally best perused by those exclusively masculine figures, the scientist and priest. It is Ida's innate recognition of this that makes her ultimately—once "Her falser self" has "slipt from her like a robe" (VII, 146)—worthy of her royal birth right and suitable to be the bride of a prince.

Vestiges, of course, was published anonymously, and as James Secord shows, "Three years after the book appeared, *Punch* could still make a great play of the enigma" by depicting it in "its number for 11 December 1847" (the year and month in which *The Princess* was first published) as a "weeping" orphan at the "entrance to a foundling hospital" with the caption, 'The Book That Goes A-begging'.[39] The true identity of the author did not come to light formally until 1884. However, early on there was speculation that the book had been written by a woman, and accusations of female authorship were used to undermine the work's credibility.[40] Harriet Martineau and Ada Lovelace were early suspects, and critics "could attribute any weaknesses to the innate qualities of the female mind in such women ... an impetuous longing after certainty [for example] made *Vestiges* just the sort of synthesis a woman might attempt".[41] Sedgwick addressed the question directly in his review, asking "But who is the author?" He came to believe it was the work of a man. However, before he made this clear, and in order to undermine the book and its author, he read *Vestiges*'s failings as the mark of a woman's authorship. Thus, he writes; at first he was led to

---

[39] Secord (2003), 22.

[40] Ibid., 20. Secord writes; "Implicit codes of propriety also meant that genteel women, whose social positions were independent of their intellectual accomplishments, could discuss the implications of *Vestiges* more readily than could some British Association lions. While Mantell, Bunbury, and Brown were condemning shoddy facts, Florence Nightingale and Lady Ashburton were speculating about the future of humanity" (416).

[41] Ibid., 20–1.

the conclusion that it was the work of a woman "by certain charms of writing—by the popularity of the work—by its ready boundings over the fences of the tree of knowledge, and its utter neglect of the narrow and thorny entrance by which we may lawfully approach it". The book displayed a "sincerity of faith and love" in whatever "system" the author has taken to their "bosom". Sedgwick thought "no *man* could write so much about natural science without having dipped below the surface, at least in some department of it". Thus, he writes, "We thought, when we began to read 'The Vestiges,' that we could trace therein the markings of a woman's foot."[42] Such textual impressions, he insinuates, would be shallow, slight, intrinsically empty and easily erased.

Sedgwick's ruse was designed to weaken *Vestiges*'s argument and to demean the male author, but it served also to remind women of their fallibility in casting judgement on such knowledge. Stringing out the ruse, Sedgwick duly apologised to women for his assumptions: "In thinking this, we now believe we were mistaken." Thus, he begins his rhetoric of appeasement; "But let us not be misunderstood. Within all the becoming bonds of homage, we would do honour to the softer sex little short of adoration." While the female mind cannot plumb the depth of science, she rules her own sphere:

> In taste, and in sentiment, and instinctive knowledge of what is right and good—in discrimination of human character, and what is most befitting in all the moral duties of common life—in everything which forms, not merely the grace and ornament, but is the cementing principle and bond of all that is most exalted and delightful in society ... But we know, by long experience, that the ascent up the hill of science is rugged and thorny, and ill-fitted for the draperies of a petticoat; and ways must be passed over which are toilsome to the body, and sometimes loathsome to the senses.[43]

Like the many "light" feet (III, 340) of the female geologists in *The Princess*, on the path towards scientific knowledge, woman cannot hope to leave a lasting footprint. And Sedgwick's metaphor of the "hill of science" also seems apt for the poem's geological expedition, as the women "Set forth to climb" with "Many a light foot" the "dark crag" and "About the cliffs" (III, 336–42). Fittingly, it is the Prince who warns

---

[42] Sedgwick, "Vestiges" (1845), 2.

[43] Ibid., 2.

Ida that her attempt to establish a solid body of female knowledge "May only make" a "footprint upon sand" (III, 223). Thus, he turns Ida's thoughts from the "even pedestal" with man, with its cultural implications, to the fossil footprint. Speculating on geological time and on what might lead to a lasting memorial of her great deeds, Ida retorts:

> Would, indeed, we had been,
> In lieu of many mortal flies, a race
> Of giants living, each, a thousand years,
> That we might see our own work out, and watch
> The sandy footprint harden into stone. (III, 250–4)

Thus, Ida takes up the "footprint" metaphor which was originally the Prince's, leading her to envisage her feminist legacy in terms of her foot marks "hardened into stone". But the markings of a woman's foot, in Sedgwick's language, specifically denote the shallowness, slightness and insubstantiality of the female mind. The footprint which is meant to signify the enduring legacy of Ida's feminism, conversely conflates it with certain kinds of geological remains, which, in turn, lead to interesting gender implications based on the various kinds of fossil evidence that were available and discussed in the 1830s and 1840s. Fossil footprints offered startling and provocative evidence for contemporary geologists about the nature of prehistoric life. However, they provided a particular type of fossil evidence, entirely different from the hard evidence of fossilised bones—those vast bulks that are the silent, incontestable signifiers of massive power and energy.

Ida's Fossil Footprint

The first papers describing fossil footprints were published in 1828 by Henry Duncan, a Church of Scotland minister and his friend, James Grierson, on the impressions found in a quarry at Corncockle Muir in Dumfriesshire.[44] Duncan writes of "The remarkable phenomenon which I am about to describe ... that of numerous impressions ... so close a resemblance to the foot-prints of quadrupeds, as to leave no doubt respecting their identity". The find was "so extraordinary" that Duncan

---

[44]For more on Duncan, see, Martin J. Rudwick, *Worlds Before Adam* (2008), 151–3.

believed it to be "unique" and to have "not hitherto attracted the share of public attention which it deserves".[45] In "Footsteps Before the Flood", also of 1828, Grierson described the same footprints as "presenting exactly such an appearance as is sometimes exhibited on the sea-beach, where the tide has passed over and filled up the footsteps of a recent traveller".[46] By 1836, the authenticity of the fossil footprints had been established. Footprints were not only evidence of the past existence of ancient life, they offered specific information about the creature who made them; they could be read to determine the length of the stride, the size and weight of the animal and even its type. Initially, Duncan described the footprints as varying from "the size of a hare's paw to that of a foal's hoof".[47] Buckland designed the "piecrust" experiment, quoted above, to establish exactly what kind of animals were responsible for the tracks and, having originally thought they may have belonged to an ancient crocodile, concluded that they were, "too short for the feet of Crocodiles, or any other known Saurians". Rather, he wrote, "it is to the Testudinata or Tortoises, that we look, with most probability of finding the species to which their origin is due".[48] Buckland remarked on the apparent ease with which such preservation might take place: "The only essential condition of such preservation being, that ... [the footprints] should have become covered with a further deposit of earthy matter before they were obliterated by any succeeding agitations of the water".[49] Chambers, much later in *Vestiges*, also drew attention to the Corncockle Muir fossil footprints: "the vestiges of an animal supposed to have been a tortoise are distinctly traced up and down the slope, as if the creature had had occasion to pass backward and forwards in that direction only, possibly in its daily visits to the sea".[50]

---

[45] Henry Duncan, "An Account of the Tracks and Footmarks of Animals found impressed on Sandstone in the Quarry of Corncockle Muir, in Dumfriesshire" (1828), 195.

[46] James Grierson "On Footsteps Before the Flood in a Specimen of Red Sandstone" (1828), 132.

[47] Duncan, "An Account" (1828), 195.

[48] William Buckland, *Geology and Mineralogy Considered with Reference to Natural Theology* (1836), I, 262.

[49] Ibid., I, 260.

[50] Chambers, *Vestiges* (1994), 101–2.

Shortly after these first examples the American geologist, Edward Hitchcock, found more footprints in Connecticut. Buckland noted that "Ornithichnites, or foot-marks of several extinct species of birds, [have been] found in the New Red sandstone of the Valley of Connecticut." They are, he writes, "so distinct, that he [Hitchcock] considers them to be made by as many different species, if not genera, of birds".[51] Chambers also notes this find: "If geologists shall ultimately give the approbation to the inferences made from recent discoveries in America, we shall have the addition of perfect birds, though probably of a low type, to the animal forms of this era [new red sandstone]."[52] In 1842, while on his travels in North America, Lyell visited the banks of the Connecticut River to see for himself "the celebrated foot-prints of birds ... beautifully exhibited."[53] He reported the "distinct footmarks of birds in regular sequence" which "faithfully" represent "in their general appearance the smaller class of Ornithicnites of high antiquity."[54] Later discoveries reported by Chambers were of footprints, which, "having a resemblance to an impression of a swelled human hand", were consigned to the "batrachian" type, and still others to "a web-footed animal of small size".[55] Of the Connecticut footprints, Buckland wrote (and Chambers also quotes this passage in *Vestiges*): the footprints are "often found crossing one another; they are sometimes crowded like impressions of feet on the muddy shores of a stream, or pond, where Ducks or Geese resort".[56] "On other tracks" Buckland reported, "the steps are shorter, and the smallest impression indicates a foot but one inch long, with a step of three to five inches".[57]

As remarkable as it was to find such striking evidence of the forgotten worlds of the distant past, these first scientifically recorded fossil footprints re-animated a world not of the gigantic and powerful 'monsters' that Richard Owen would later dub 'dinosaurs', but a rather more familiar world of docile animals, of hares, foals, tortoises, ducks, geese

---

[51] Buckland, *Geology and Mineralogy* (1836), II, 39.

[52] Chambers, *Vestiges* (1994), 103.

[53] Charles Lyell, *Travels in North America* (1845), I, 252.

[54] Ibid., II, 168.

[55] Chambers, *Vestiges* (1994), 102.

[56] Buckland, *Geology and Mineralogy* (1836), II, 39.

[57] Ibid., II, 40.

and frogs. If Ida's hope is to see her feminist legacy "harden" like a fossil "footprint ... into stone", and to "plant a solid foot into the Time" (V, 405), then her metaphor is an unfortunate one. The discoveries and discussions around fossil footprints conjured something much less monumental or prodigious than the fossil "bones" of the "vast bulk", which, "struck out" of the "black block" (III, 274–7) of the distant past, intrude upon the present as a reminder of the perennial sovereignty of brute force. Fossil footprints, conversely, seem to animate a world no less perennial but one of small, innocuous, docile and sociable creatures, most often birds, tramping backwards and forwards or in and around the much more familiar and domesticated environs of the duck pond. Indeed, the association with women and birds in the poem has already been made. In further examples, Melissa is said to be "light, / As flies the shadow of a bird" (III, 80), the men (in order to pass off as women) take on "maiden plumes" (I, 199) and for the Prince, Ida herself is an eagle (III, 90). The flocking instinct is significant here too, as the young women students left alone are docile, sociable creatures, safe in the association of their own kind: they are "like morning doves / That sun their milky bosoms on the thatch" (II, 84–5). However, once disturbed—as when the new students are discovered to be men—they soon become flustered; "A troop of snowy doves athwart the dusk, / When some one batters at the dovecote-doors, / Disorderly the women" (IV, 150–2). Here, the women, no longer imbued with the mythical stature of the 'siren' or the winged "woman-statue", display instead the skittish behaviour of birds 'in a flap'. With the armies of the North assembling at the palace gates, the women "Fluctuate[d]" "to and fro" (IV, 461–460), or they become like "wild birds" who panic-stricken "Dash themselves dead" on the "Fixt ... beacon-tower" (VI, 472) of Ida's mythic nobility. Ida is able to quell the frenzy; stretching out her arms she asks: "What fear ye, brawlers? am not I your Head?" (VI, 477)—yet clearly the inference is that without their "Head", the women will act as proverbial chickens do.

Not all the fossil footprints discussed, however, belonged to small birds. Geologists also noted larger ones among them. Buckland, for example, writes: "The most remarkable among these footsteps, are those of a gigantic bird, twice the size of an Ostrich".[58] Chambers, on the

---

[58] Ibid., II, 39. It was not known at this time that these were the tracks of an early saurian.

same evidence, notes that "One animal, having a foot fifteen inches in length (one-half more than that of the ostrich,) and a stride of from four to six feet, has been appropriately entitled, *ornithichnites giganteus.*"[59] Clearly, not all birds are equal, and this gigantic bird is made all the more fearsome in the conflation of great size and the limited intelligence associated with birds. In a poem in which size really does matter, Ida is also distinguished by her size. When she rises "Once more through all her height, and o'er him [Cyril] grew / Tall as a figure lengthened on the sand" (VI, 144–5), she proves herself a fabulous bird capable of leaving a striking impression not only on the men but also, as her shadow "lengthens", on the sands of time. What might be read from her impression, however, is less than heroic. Like the gigantic bird of prehistory, Ida is monstrous, as she acknowledges herself: "No doubt we seem a kind of monster to you" (III, 259). The bird-brain tyrant is the worst of all tyrants, and women (merely 'big children' in Schopenhauer's notorious mid-century view) were deemed ill-equipped by nature to exert the reasoning powers associated with male adulthood. Where "Many a little hand" and "Many a light foot" (III, 338, 340) take the 'dip' of the 'strata', Ida is distinguished by her size. She has "long hands" and she impresses when she rises to her full "height" (II, 26–7). She stands "Among her maidens, higher by the head" (III, 163), and she must descend from her "fixt height" (IV, 289) to become part of the "milky rabble of womankind" (IV, 290). Ida, as her brother states, "flies too high" (V, 271); she is exceedingly more impressive than any other woman in the poem, but she is misguided and in her unruly state she is a tyrant; she is Lilia's childish petulance grown solemn, grave and unbending. The Prince indicates that Ida is exceptional among women when he introduces the metaphor of the footprint in the sand. What if, he asks, a "feebler heiress ... ruins all"? (III, 221). As Ida is a fabulous fiction, the implication is that there can be no able "heiress" to lead women's emancipation after Ida. Beyond the framed narrative of the fairy tale there are only ordinary, real women—who are limited by their biology, such as the 'little' Lilias, whose youthful exuberance needs directing safely into the domestic space where she can find release for her energies in the ministries of the home and where her authority extends no further than the nursery walls.

---

[59] Chambers, *Vestiges* (1844), 104.

The gigantic bird who left impressions of its feet as fossil prints in the sand of some primordial beach, and who once stood at the head of some ancient avian pecking order, is now itself extinct. And similarly, Ida, who is larger than life, like the 'warrior lady' of the chronicle, does not exist. As Walter teasingly asks Lilia in the Prologue, with reference to the warrior lady, "Where … lives there such a woman now?" (124–6). To counter this, Lilia mobilises arguments about social determination: "There are thousands now / Such women, but convention beats them down: / It is but bringing up; no more than that" (127–9). And yet, these arguments are undercut by the text's repeated emphasis on the ultimate superiority of size and physical strength; Sir Walter, after all, as an example of benevolent nobility, is "No little lily-handed Baronet" but "A great broad-shouldered genial Englishman" (Conclusion, 84–5). The hard facts are that, as Lyell had suggested, "In the universal struggle for existence, the right of the strongest eventually prevails" (*PG*, II, 56). Herbert Tucker demonstrates how the threat of physical violence is ever-present in the poem.[60] Every concession to Ida's feminism is ultimately dependent on the limits of male tolerance. The Prince's gentlemanly forbearance is part of the threat, as when the armies of the North and South assemble, it becomes clear that Ida's continued resistance to his amorous, yet chivalrous, advances is exactly what makes violence inevitable. The female university is always liable to be crushed by the superior physical forces of the male armies. To emphasise the importance of size, Lilia's name is consistently prefixed by "little"; she is "A rosebud set with little wilful thorns", "The little hearth-flower Lilia" (Prologue, 152, 165), "little Lilia" (Conclusion, 12, 116), and the homesick boys are like "many little trifling Lilias" (Prologues, 186). The little Lilias of the nineteenth century do not have the warrior lady's "stature more than mortal" (Prologue, 40). Neither do they have Ida's impressive stature nor her royal prerogative. Thus, it is not "convention" that keeps women subordinate in the text's logic but nature—the natural limits of their intelligence combined with the inferiority of their physical strength.

Fittingly, in response to Walter's teasing, Lilia taps a "tiny silken-sandaled foot" (Prologue, 149). The text raises the question: what kind of impression could be left by Lilia's tiny foot, or the "light foot" (III,

---

[60] Tucker, *Tennyson* (1988), 353–7.

340) of the women geologists, or the "little-footed" (II, 118) Chinese, or even Ida's "tender foot, light as on air"? (VI, 73).[61] Only, perhaps, footprints that recall the world of women "Whose brains" as Ida suggests, "are in their hands and in their heels"; who are "But fit to flaunt, to dress, to dance, to thrum, / To tramp, to scream, to burnish, and to scour" (IV, 498–9). Pointedly, the ideal model of womanhood in the poem is, of course, the Prince's mother, who was:

> Not learnèd, save in gracious household ways,
> Not perfect, nay, but full of tender wants,
> No Angel, but a dearer being, all dipt
> In Angel instincts, breathing Paradise,
> Interpreter between the Gods and men
> Who looked all native to her place, and yet
> On tiptoe seemed to touch upon a sphere
> Too gross to tread. (VII, 299–306)

The mother is conscious of the biology that anchors her to her natural sphere, and thus, she walks "On tiptoe", as if to "touch" what she is otherwise "Too gross to tread" (VII, 305–7), and in walking on tiptoe, she strives both to live up to an ideal of womanhood, while endeavouring to leave no footprint at all. If Sedgwick thought he could "trace" in *Vestiges* the "markings of a woman's foot", he suggests that he sees the trace of superficiality and of a lack of rigour—the markings, in fact, of the bird-brain, of the "falcon-eyed" (II, 88) Psyche, whose "bird's-eye-view" of history, all too broadly "Glance[s]" (II, 107), with no eye for detail, and picks out only what it likes. The women, Cyril says, "hunt old trails" (II, 368), their footprints, like the fossil footprints discovered by geologist in the 1830s and 1840s, leave no impression of their mastery over their world; rather, they indicate the toing and froing of docile, domesticated feet over the same quotidian ground.

---

[61] Zimmerman reads the footprint more optimistically, arguing that "Ida's efforts and those of her comrades leave traces; they will, in time, offer women a better life. No catastrophe will come with revolution in its wake; instead, women must look to the power of a single drop of water to wear down a mighty stone over many years" (*Excavating Victorians*), 83.

If the women in the story 'flock' like birds, the allusion to sheep must also be noticed.[62] When the sanctuary of the university is under threat, the women gather "thick as herded ewes" (IV, 458), exhibiting herding instincts that suggest their homogeneity. Ideas around women's lack of individuality are vocalised unambiguously by the Prince's father, who frames sexual relations in the language of hunting and herding. Man, like a herdsman, "leaps in / Among the women, snares them by the score / Flattered and flustered" (V, 155–7). Repudiating his father's generalised view of women, the Prince charges him with "clash[ing] them all in one", when they "have as many differences as we. / The violet varies from the lily as far / As oak from elm" (V, 172–5). Paul Turner argues that Tennyson attempts to differentiate between female types, reading these lines as influenced by Mary Wollstonecraft's complaint that "though men are allowed to be individuals, 'all women are to be levelled, by meekness and docility, into one character of yielding softness and gentle compliance'".[63] However, the blithe metaphor of flowers/women and trees/men hardly lessens the view of women as collectively meek, docile, yielding, gentle and compliant. The hopeless homogeneity of women as "herded ewes" is one of the lessons that Lilia must learn from the story of Ida. As Lilia imagines an all-women university, she is reminded by Walter of the physical attributes that she shares with all young women as he imagines a university containing "many Lilias" (Prologue, 146). Lilia is levelled by her physical femininity; the attributes that the text suggests she shares with all young women and that are consistently foregrounded as the most valuable feminine attributes: the "golden hair" of the "sweet girl-graduates" (Prologue, 42), the "gemlike eyes" and "golden heads" (IV, 459–60) of the "maids" gathered in fright as the summer palace is besieged by armies. Ida too must surrender her individuality. She tells the Prince, "You cannot love me" (VII, 317), acknowledging that it is not her he loves but an image of femininity to which she must conform; the ideal that he insists she becomes.

---

[62] Reading section IV, lines 448–68, Herbert Tucker draws attention to the "extraordinary mix of concessions to the pleasure principle and to the reality principle: the former, because it caters to cultural ideals of feminine loveliness (snowy shoulders, gemlike eyes, open mouths); the latter, because it represents the effect such ideals have in foredooming the kind of radical reform Ida has sought" (*Tennyson*, 1988), 360.

[63] Paul Turner, *Tennyson* (1976), 111.

The herding, animalistic characterisation of women draws on mid-century gendering that saw man and woman as "The single pure and perfect animal" (VII, 288). As the old king Gamma says in what are perhaps the most-quoted lines of the poem:

> Man for the field and woman for the hearth:
> Man for the sword and for the needle she:
> Man with the head and woman with the heart:
> Man to command and woman to obey;
> All else confusion. (V, 436–40)

This law, he says "is fixt" (V, 434), and while the Prince protests that hunting women like prey may not be the best way to win them, the final resolution of Ida's story indicates that it is the idea of the "hunt" that is objected to but not the gender roles themselves, which are fixed and not at issue in the poem's rhetoric. Such fixity leads back to Miller's non-progressive view of humankind. Appropriating Miller's geology to fit the poem's gender politics, the suggestion is that any possibility of the evolution of women out of their "fixt" relationship to men would disrupt the natural order and lead to social discord—to Gamma's utter "confusion". This threat is one that Lyell, Miller and Sedgwick all recognise in their various ways, in terms of the effect of extrapolating evolutionary analogies across the power structures of social class and gender. Lyell's reaction to Lamarckian evolution is a good gauge of the insidious threat that evolution posed to social organisation. Arguing against developmental theories and Lamarckian evolution, Lyell was at pains to suggest that while there could be a certain amount of flexibility within species boundaries, there could be no transmutation between species. Refuting (as Miller does later) what appeared to be progressive complexity in the fossil record, Lyell addressed the relatively recent emergence of man:

> can this introduction be considered as one step in a progressive system by which, as some suppose, the organic world advanced slowly from a more simple to a more perfect state? To this question we may reply, that the superiority of man depends not on those faculties and attributes which he shares in common with the inferior animals, but on his reason by which he is distinguished from them. (*PG*, 1, 155)

If man had not the capacity for reasoning and had only the instincts asso-ciated with "lower animals", Lyell writes, "he might then be supposed to be a link in a progressive chain" (*PG*, I, 155). In the simplest terms, the superiority of man is based on his reasoning ability. If the "single pure and perfect animal" of man relies (paraphrasing Ruskin) on each sex supplying what the other lacks as "each fulfils / [the] Defect in each" (VII, 285–6), then men supply reason, while women, "truer to the law within" (V, 181), supply the natural instinct that ensures reproduction unfolds successfully—the "bearing and the training of a child" being "woman's wisdom" (V, 455–6). In this crude dichotomy, women are fixed by the demands of a set of relations which operate to sustain the species, and any hope of progression out of their fixed role is a rejection of nature itself—a 'confusion' leading, in its extreme, to extinction itself.

The absurdity of the conflation of women and knowledge that the university represents was obvious to a majority of readers immersed in contemporary gender ideology. Samuel Dawson, as already suggested, understood this to be the nub of the text's comedy; "It is the incon-gruity of opposing function which excites laughter", as seen in the jux-taposition of nature and learning embodied in Psyche—"that lactiferous Doctor of Philosophy".[64] The nursing mother and the doctor exist in opposition and in mutually exclusive realms described in the strict nomenclature of species classification. As Londa Schiebinger argues, Linnaeus introduced the term *Homo sapiens* to discriminate between the 'man of wisdom' and other primates. Man, she writes, "had traditionally been distinguished from animals by his reason; the medieval apposition, *animal rationale*, proclaimed his uniqueness". However, his coinage of the term *Mammalia*, shows how "in Linnaean terminology, a female characteristic (the lactating mamma) ties humans to brutes, while a tra-ditionally male characteristic (reason) marks our separateness".[65] Woman is "not undevelopt man" (VII, 259), she is an element of man that links him to the biological. Thus, there can never be equality between men and women, as only in the union of their different elements can they make the perfect whole that is mankind. Arguments for equality are the "foolish work / Of Fancy" (VI, 97–8) that proves the point. Thus, just

---

[64] Dawson, *A Study* (1882), 22.

[65] Londa Schiebinger, "Why Mammals are Called Mammals: Gender Politics in Eighteenth-Century Natural History" (1998), 394.

as "dear Lady Psyche" must learn "that the mother-hunger cannot be appeased by primal nebulæ!", Ida must learn that her hope for intellectual equality is "Beyond all reason" (I, 142) because reason itself is beyond women.[66]

If reason is the realm of men, the women in the poem, including Ida, demonstrate their faulty reasoning by attempting to usurp the male role. All attempts by the women to occupy intellectual spheres are thwarted by their natures, from Lilia's immoderate death penalty in the Prologue, Psyche's surrender to the domestic, to Ida's kidnapping of Psyche's child. The women must learn to be shepherded, as the Prince tells Ida: "the children call and I / Thy shepherd pipe" (VII, 202–3). Their herding instinct allows the women to be led, and it is natural for animals that herd to follow a lead. As Lyell suggests in his argument for the fixity of the boundaries of species, "An animal in domesticity ... is not essentially in a different situation in regard to the feeling of restraint from one left to itself ... There is nothing in its new situation that is not conformable to its propensities; it is satisfying its wants by submission to a master, and makes no sacrifice of its natural inclinations" (*PG*, II, 42–3). The animal with an instinct to herd will always "obey some individual, which by its superiority has become the chief of the herd", while "no solitary species, however easy it is to *tame* it, has yet afforded true domestic races" (*PG*, II, 43). Quoting M.F. Cuvier, Lyell argues that herd leaders are "distinguished in the general mass by the authority which they have been enabled to assume from their superiority of intellect", and likewise "every social animal [who] recognizes man as a member, and as the chief of its herd, is a domestic animal. It might even be said that from the moment when such an animal admits man as a member of its society, it is domesticated, as man could not enter into such a society without becoming the chief of it" (*PG*, II, 43–4). In the dichotomy of gender traits and gender roles which subsumes women into the domestic category, women— weaker in reason and complying with their natural propensity to herd and to follow—must naturally concede to men. Ida is only 'Head' while the single-sex institution is sustained; once men enter, the natural order (the order inscribed by Sir Ralph's statue and the remains of "master existences") falls back into place. As Ida reflects, "I should have had to do with none but maids / That have no links with men" (VI, 273–4).

---

[66] Dawson, *A Study* (1882), 44.

Lyell ends by suggesting, "It seems reasonable to conclude, that the power bestowed on the horse, the dog, the ox, the sheep, the cat, and many species of domestic fowls, of supporting almost every climate, was given expressly to enable them to follow man throughout all parts of the globe—in order that we might obtain their services and they our protection" (*PG*, II. 44). The relationship of man to those animals predestined to be domesticated is perfectly in harmony with man's relationship to woman. Reason marks the distinction between men and animals and between men and women, and women's reduced reasoning capacity, like those of domestic animals, is what enables them to support their subordination. Such gendering is written into the poem's form; the narrative is the men's, while the women, bird-like, sing the refrains. The plot of their story, like the inscription on the winged "woman-statue", is "too deep in shadow"—too deeply ingrained in their biology to be articulated in the rarefied realms of reason. Their song plays out against the "clocks and chimes" of biological time (the linear and generational time of *The Winter's Tale*), which (as the young men enter the world of women) threatens to drown out the male grand narrative as they "scarce" can "hear each other speak". Responding to innate impulses, the women sing, and, like the caged "nightingale, / Rapt in her song", when they are true to their natures they are "careless of the snare" (I, 210–8).

IDA'S 'TRUE' REMAINS

While Sir Ralph's statue and the fossil bones gesture towards non-linguistic epistemologies, they nevertheless exert a dynamic power to mean and to generate narratives. Buckland made exciting claims for the power of remains as offering the "great master key whereby we may unlock the secret history of the earth". Fossil remains are "documents which contain the evidences of revolutions and catastrophes, long antecedent to the creation of the human race; they open the book of nature, and swell the volumes of science".[67] Miller, equally awed by the door that fossil remains open onto the past, suggests that the contemplation of "remains is a powerful stimulant to thought. The wonders of Geology exercise every faculty of the mind—reason, memory, imagination" (*ORS*, 101). The fossil footprint, however, is a troublesome indicator of the past but

---

[67] Buckland, *Geology and Mineralogy* (1836), I, 128.

at the same time, an apt indicator of the simultaneous presence and non-presence of women in the story of the human species. As a metaphor for the continuing influence of Ida's feminism, the fossil footprint works precisely to foreground the insubstantial nature of her legacy; it represents "the markings of a woman's foot". The fossil imprint is negative evidence of existence; it is an empty signifier, while Sir Ralph's statue and the bones of "master existences" offer hard, material evidence of dynamic existence. The empty signifier is foregrounded in the framing narrative too, in Lilia's "empty glove upon the tomb" of Sir Ralph, which offers merely a "model of her hand" (V, IV, 20–1). This token of vacuity enters the framed story as the favour for which the Prince must fight. It is the "model" of femininity that signifies nothing substantial at all, and nothing outside its own flesh.

Finally, Ida's "falser self slipt from her like a robe", leaving "her woman, lovelier in her mood / Than in her mould" (VII, 146–8). She concedes "In sweet humility" that she has "failed in all; / That all her labour was but as a block / Left in the quarry" (VII, 215–7). Unable to carve her monumental legacy into the unhewn rock, Ida crudely remains undifferentiated from nature. There can be no precedent in the past for the equality of women, and Ida can leave no solid remains to signify her singular legacy for future generations of women. Thus, the poem accomplishes the defeat of feminism past, present and future—once and for all. Bereft of purpose, Ida must find meaning elsewhere. She fixes her identity to another form of remains—to the softer, more intimate remains suggested by the lock of her hair which the Prince has possessed since Ida's mother ("ere the days of Lady Blanche") "shore the tress" (VI, 94, 98). The lock reminds Ida of the biological legacy that she herself represents as her mother's daughter. Like all remains, the lock of hair can be read, and it offers a particular narrative—not one that speaks of "great deeds" or "master existences" but one that tells of filial and domestic relations. The lock of hair encodes the natural woman, as Ida—unlike Lady Blanche, whose "autumn tresses", we are pointedly told, are "falsely brown" (II, 426)—is still possessed of her natural self. Fixing her significance to the filial nexus, Ida's "dark tress" has an erotic quality too. It metonymically stands for Ida's sensual, physical presence, and like her picture, it elicits voluptuous narratives as "Sweet thoughts ... swarm as bees about their queen" (I, 38–9). The lock of hair circumscribes Ida's sphere of meaning. However, while it is meaningful to the loved one who gazes on it, it is meaningless outside the realms of intimacy. In such remains, Ida's value originates in the private gaze; her significance and her identity are not durable but contingent on the viewer as she becomes the reflection of his gaze. More profoundly, however, the lock of hair confronts Ida with her own remains; as Deborah Lutz suggests, "numerous ideas about death and the body accreted in and around these little things". The lock of hair "affirms the death written in the bodies while they are still animated".[68] Ida must resign her hope to leave a monumental legacy and devote herself to ensuring a biological and individually unspecific legacy—the soon-to-be-forgotten legacy of the lock of hair.

At the end of the poem, the Prince offers a self-serving narrative concerning the progression of women's rights. However, any progress is reliant on Ida relinquishing the power to direct that progress and submitting to the fixed authority of the past as it is written in remains. Thus, *The Princess* works hard to figure a progress for women that actually amounts to no progress at all. It takes Miller's insistence on the non-progressive state of nature to make a gender ideology of irresolvable difference appear natural. Geological remains are mobilised to indicate how nature's laws are fixed. Thus, there can be no change in the relationship between the sexes—no catching up with men for women, for "could we make her as the man, / Sweet Love were slain" (VII, 260–1)—presumably resulting in the extinction of the species. Returning to the framing narrative, Lilia, still dissatisfied, turns to the maiden Aunt and demands: "You—tell us what we are." However, she "who might have told" cannot tell because the Aunt, "crammed with theories out of books" (Conclusion, 34–5), has lost her connection to her natural self. Aunt Elizabeth cannot answer Lilia's question, and, as the day closes on her theories, the parable of Princess Ida has made them redundant; nature overtakes her, as the sun sets on Lilia's feminist outburst. Divesting Sir Ralph's statue of its female garb, Lilia is now silenced, and must reluctantly succumb to the implacable evidence of Sir Ralph's image set in stone. Dressing up the stone statue in women's clothes will not rewrite history and will not alter the hard, material, geological facts of nature. As Ida must be disrobed of her 'mannish' self, so Sir Ralph is disrobed of female attire once the gender order is restored. The statue, like the fossil remains of the "vast bulk" that lived "before man was", is material evidence that tells of the strictly gendered order of serial

---

[68] Deborah Lutz, *Relics of Death in Victorian Literature and Culture* (2015), 130, 138.

creation and enshrines the inviolable passage of power through geological time. Beyond their biological offspring, women, however, leave no mark of individual power or substance. The woman's textual foot *print* (as Sedgwick saw it) leaves a trifling mark soon erased, while Ida's sandy footprint hardened into stone points merely to the tramp of unremarkable, docile feet (as suggested by geological evidence). The fossil footprint is entirely appropriate for signifying Ida's presence/non-presence, for her gender dictates that she cannot leave behind the evidence of her individual self and her great deeds. Instead, she leaves a much more fitting emblem of her existence: the empty space—she leaves, in fact, nothing intrinsically of herself at all.

CHAPTER 3

"Uniformitarian Arguments Are Negative Only": Lyell and Whewell

"In the economy of the world", said the Scotch geologist [James Hutton], "I can find no traces of a beginning, no prospect of an end"; and the declaration was the more startling when coupled with the doctrine, that all past changes on the globe had been brought about by the slow agency of existing causes. (Charles Lyell, *Principles of Geology*)[1]

The Catastrophist is affirmative, the Uniformitarian is negative in his assertions: the former is constantly attempting to construct a theory; the latter delights in demolishing all theories. The one is constantly bringing fresh evidence of some great past event, or series of events, of a striking and definite kind; his antagonist is at every step explaining away the evidence, and showing that it proves nothing. (William Whewell)[2]

Charles Lyell's three-volume *Principles of Geology* (1830–1833), as a work under constant revision, went through twelve editions. While the uniformitarian principles it expounded remained always the same, the text itself was continually revised, refined and re-stated over the forty years from its first publication to Lyell's death in 1875—the final edition being published posthumously. Lyell it seems could not let go of his work and thus, perhaps fittingly for a text postulating an earth endlessly

---

[1] Charles Lyell, *Principles of Geology*, 3 vols., 1830–1833 (1990), 1: 63. Hereafter cited as *PG* parenthetically by volume and page number.

[2] William Whewell, *Indications* (1845), 155.

© The Author(s) 2017

M. Geric, *Tennyson and Geology*, Palgrave Studies in Literature, Science and Medicine, DOI 10.1007/978-3-319-66110-0_3

forming and re-forming itself, the text (for Lyell at least) never really achieves closure. Tennyson struggled with closure too, as *In Memoriam* and *Maud* both attest. Both are poems modelled on a Lyellian way of seeing; *In Memoriam* struggles to make linear progress, while *Maud*'s characters shift between human and deep time. This chapter examines the Lyellian geological ideas that inform *In Memoriam* and *Maud*. *The Princess*, as Chap. 2 argued, harnessed the ideological patterns at work in Hugh Miller's geology, while for *In Memoriam* and *Maud* ideology is dissolved as the poems take on and interrogate the radical potential of Lyell's geology. *Principles* set out to forge a particular form of writing about science that avoided the teleological assumptions that characterised writing on natural phenomena in the period. Lyell's approach looked back to Enlightenment traditions in its empirical tone but was recognisably a different type of writing on science partly because of the limits it placed on broad speculations and its emphasis instead on observable data. *Principles* set out to define the disciplinary borders of geology and it therefore marked an important moment in the history of geology's professionalisation. The tension between, on the one hand, *Principles'* specialisation, and on the other, transdisciplinary natural philosophy can be seen in the comparison between Lyell's approach and that of the philosopher of science, William Whewell. As discussed in Chap. 1, Whewell's polymathic mentality was part and parcel of his adherence to a teleological perception of the unity of knowledge, what he termed the "consilience of inductions" (the 'jumping together' of inductions), which describes his conviction that empirical observation leads to inductions that in turn reveal 'general truths'.[3]

This chapter examines the rhetorical strategies Lyell employed in writing *Principles*, along with Whewell's critique of Lyell in order to demonstrate that Tennyson's poetics responded to and fed into the rhetoric of both men. The chapter explores the differences in Lyell's and Whewell's mentalities and reads these differences (in Chaps. 4 and 6) as worked through in *In Memoriam* and *Maud*. These poems are structured on Lyell's science; they take Lyell's reasoning, the contradictions in his rhetoric, to their logical conclusion, and in doing so, they not only expose

---

[3] The concept is explained in full in volume two of Whewell's, *History of the Inductive Sciences*, 3 vols., 1837 (1967). For a comprehensive analysis of Whewell's 'consilience of inductions', see Larry Laudan, *Science and Hypothesis* (2010).

Lyell's ideological manoeuvring, but textually perform the wider episte-mological implications of uniformitarianism. Lyell's "history of geology" that prefaces the first volume of *Principles* is examined to demonstrate how Lyell uses uniformitarianism rhetorically as well as geologically, to establish his approach as a rational, 'scientific' approach—a new 'uniform-itarian' way of thinking. Whewell's writings are seminal in understand-ing Lyell as well as Tennyson's uniformitarian poetics, and his key texts, *History of the Inductive Sciences* (1837) and *Indications of the Creator* (1845), which engage directly with *Principles*, are also examined. The chapter begins, however, by addressing some of the issues involved in reading 'science' and 'literature' together and examining the misreadings and misinterpretations of geological terms and ideas, finally looking spe-cifically at the term 'uniformitarianism', as its meaning and use is crucial for this study as a whole.

SCIENCE AND LITERATURE

In terms of the current critical debates around reading science and lit-erature together, a practical point about the cultural reading of science needs to be made, as where the history of science and literary criticism intersect there have been issues about the interpretation of scientific ideas. Lyall I. Anderson and Michael A. Taylor, in their valuable exami-nation of Tennyson and geology (2016) from the perspective of the his-tory of geology, suggest that cultural interpretations of Lyell's geology have sometimes been misleading. Lyell's theory of climate change, for example, has been read as suggesting that he saw geological change as cyclical. Taylor and Anderson suggest, "despite what is sometimes said, Lyell's analysis prescribed no particular pattern, linear, cyclical or oth-erwise, to future climate change, which depends on the vagaries of the changing distribution of land and sea".[4] James Secord also notes this misinterpretation. Lyell, Secord argues, "was not the 'historian of time's cycle'" (as Stephen Jay Gould has suggested). In fact, Secord suggests, that "many of the best modern commentators on the *Principles* ... have taken Lyell's statements about the pattern of earth history out of con-text" and not read them for what they are, "thought experiments about

---

[4]Anderson and Taylor, "Tennyson" (2015), 365–6.

the past".[5] Such misreadings, however, will always interest literary critics as they are the stuff of nuanced analysis and are highly significant in
themselves. It was, for example, the lack of attention to Lyell's rhetoric and Whewell's coinage and clever characterisation of 'uniformitarianism' that led to a century or more of misunderstanding in the history
of geology. In the case of climate change, the misinterpretation was, as
Anderson and Taylor show, "encouraged by Lyell's metaphorical reference in a literary flourish to 'what may be termed the winter of the
"great year", or geological cycle'".[6] It was precisely Lyell's misleading metaphor that Tennyson took up in a letter to Richard Monckton
Milnes. Referring to *Principles* in parenthesis he light-heartedly reflects
on the November weather in terms of a "great Geological winter", and
refers to "latent heat" that "tries ... to become sensible", the "wear[ing]
down [of] the pole", and the earth's "grow[ing] up at the equator"; all
images drawn from volume II, chapters nine and eighteen of *Principles*.[7]
Tennyson was responding to Lyell's rhetoric, his patterns of thought,
his thought experiments, and it was Tennyson's reading of Whewell's
account of *Principles* that probably further encouraged him to see Lyell's
geology as proposing cycles, as Whewell also characterised Lyell's geology as cyclical, as discussed below.

The translation of directionless change into cycles by Whewell,
Tennyson and various later literary critics and historians of science makes
the attention to language even more crucial and shows how important it
is to recognise, as far as possible, how ideas travel through language and
how ideas are misinterpreted or subverted to meet the requirements of
particular rhetorical needs (and Whewell himself is adept at noticing such
rhetorical strategies in Lyell's work). If Lyell's writings, or any other
geological writings, are misunderstood or misrepresented, purposely or
otherwise, in contemporary discourses, literary or otherwise, then the
effects of these misinterpretations and the reasons for them need attention. It seems particularly important to explore the ideas that grew up
around the language Lyell used to describe his geology. Thus, Anderson,

---

[5] James Secord, "Introduction," in *Charles Lyell, Principles of Geology*, J.A. Secord ed.
(1997), xix.

[6] Anderson and Taylor, "Tennyson" (2015), 346.

[7] A.T. to R.M.M., Nov 1 (1836) *Letters of Alfred Lord Tennyson I, 1821–1850*, Cecil Y.
Lang and Edgar F. Shannon, Jr. eds. (1982), 145. Hereafter cited as *Letters ALT* by volume and page number.

Taylor and Secord make very valid points; critical writing about literature and science in the period must demonstrate a sound understanding of the science, but one of the major reasons for doing this is to be better able to spot contemporary misinterpretations, some of which were purposely misleading. The point is to attend to the nuances of language—those times when the geologist is misinterpreted or the geologist's use of language misrepresents his own idea—as these losses or gains in translation are rich areas for analysis that lead to more authentic and subtle accounts of literature, science and culture.

READING UNIFORMITARIANISM

The problem of misinterpretation and misrepresentation has led to confusion over the meaning of the term 'uniformitarianism', particularly when used to describe Lyell's geology, partly because Lyell did not use the term himself, and partly because it has so often been used to conflate two very different ideas. The aim of Lyell's *Principles of Geology* was to "establish the *principles of reasoning*" in geology, which are "neither more or less than that *no causes whatever* have from the earliest time to which we can look back, to the present, ever acted, but those *now acting*; and that they never acted with different degrees of energy from that which they now exert".[8] This, in a nutshell, represents the view of geological change that came to be known as uniformitarianism, a term coined by Whewell in 1832. Arguing according to a uniformitarian methodology and postulating geological deep time, Lyell demonstrated how seemingly ineffectual agents, such as gradual erosion, sedimentation and uplift, acting continuously and virtually imperceptibly in the present, were responsible for the most fundamental reformations of the landscape. Lyell's uniformitarianism set out to oppose theories that suggested geological change had acted at various times in the past with greater intensity than at present or that postulated periods of violent, global geological change in the earth's past that had no equivalent in modern history. The term "catastrophism", which Whewell also coined, came to describe a range of theories often understood in opposition to uniformitarianism.

---

[8] Lyell to Roderick Murchison, January 15 (1829) *Life Letters and Journals of Sir Charles Lyell, Bart*, Katherine Lyell ed., 2 vols. (1881), I: 234.

Historians have stressed that the polarisation of these broad terms has been exaggerated and that they were not, in fact, thought to be mutually exclusive by the majority of geologists working in the early nineteenth century. Geologists were not divided between two camps—the catastrophist and the uniformitarians—and to see them as such belies what were in practice nuanced positions.[9] In reality, most geologists in the early nineteenth century worked from a basically uniformitarian premise while hypothesising catastrophe when and where observable data suggested discontinuity in the geological record. Lyell, however, came to be seen as the 'founder of modern geology' largely because of his uniformitarian methodology and, Stephen Jay Gould argues, in the history of geology that followed Lyell, "the two sides received names—catastrophism for the vanquished, uniformitarianism for the victors", names that "wrap any remaining subtlety into neat packages".[10] Early historians of geology wrote Lyell into the history of geology as almost solely responsible for the modernisation of geology, his uniformitarian vision equating to a clear-sightedness that cut through the theological fog and Mosaic mist of natural theology. That Lyell's rhetoric had such a powerful and lasting effect on the history of geology is important in itself, as that same effect worked on the early readers of Lyell, such as Tennyson. Gould suggests that the dichotomy of uniformity and catastrophism was routinely "widened" by what he calls the "later textbook cardboard" that draws a line between the two concepts, when in fact in early nineteenth-century geology there had never been "a war between uniformitarian modernists and a catastrophist old guard with a hidden theological agenda".[11]

If uniformitarianism and catastrophism have been set up as false dichotomies, the use of the term uniformitarianism is still more troublesome. "All revisionists" of the "textbook cardboard" myth of geology, Gould writes, agree on a central point:

> Lyell united under the common rubric of uniformity two different kinds of claims—a set of methodological statements about proper scientific procedure, and a group of substantive beliefs about how the world really works.

---

[9] See Nicolaas Rupke, *The Great Chain of History* (1983), 193.

[10] Stephen J. Gould, *Time's Arrow, Time's Cycle* (1987), 112.

[11] Ibid., 117.

The methodological principles were universally acclaimed by scientists, and embraced warmly by all geologists; the substantive claims were controversial and, in some cases, accepted by a few other geologists.[12]

Lyell's approach was, as he called it himself, the "doctrine of absolute uniformity", which asserted that "events in the deep past had never been of greater extent, suddenness, intensity, or violence—let alone of different kinds—than actual causes" (*PG*, 1, 87).[13] In the minds of early historians, Lyell's uniformitarianism came to be associated with an empirical and modernising approach, while catastrophism became associated with Mosaic history and timescales. The term 'catastrophism', if anything, is even more slippery, as it has been used to refer to a number of geological positions, from hypotheses that postulate serial global catastrophes, to ideas that the earth has become increasingly less subject to catastrophe. However, there was nothing particularly unscientific about the catastrophist position; most catastrophists reasoned from a basically uniformitarian position, but did not go as far as Lyell in ruling out the possibility of other types of geological change where geological features implied them. Lyell's adherence to uniform change seems unscientific by his own standards, but uniformity for Lyell was so much more than a methodology through which to examine geological change; it was, as this chapter goes on to explain, a state of mind, a superior, rational and refined style of thinking.

Adelene Buckland tackles the misreading of the history of geology that emerged from the uniformitarian/catastrophist debate, arguing that two main errors have distorted perceptions of Lyell and contemporary geologists. To begin with, Lyell "did not label his theory 'uniformitarianism' (he did not, in fact, label it a 'theory' at all, since he, like his

---

[12] Ibid., 118–9.

[13] Other more specific terms were available also, as Martin Rudwick shows, the Swiss geologist and meteorologist Jean-André de Luc (1727–1817) used the term "*actual causes*" to describe the approach that looks at processes *actually* occurring in order to help understand the operations of the past, but that does not, however, assume that "everything", in terms of geological features, can necessarily "be explained by the gradual action of ordinary processes observable in the present day". "Hence", as Rudwick writes, "the analytical term *actualism*, applied to the earth sciences, denotes the methodological strategy of using a comparison with observable present features, processes, or phenomena as the basis for inferences about the unobservable deep past: in epigrammatic form, 'the present is the key to the past'" (*Worlds Before Adam*, 2008), 14, 15, n. 4.

colleagues, was suspicious, if not of theorising in general, then at least of its liability to excess)". Secondly, "the arguments made against Lyell by his colleagues were not, as has sometimes been assumed by critics, made in order to defend the biblical account of creation, and thus of a younger earth, against Lyell's 'realist' view".[14] Buckland, of course, is right on both counts; the term uniformitarianism was projected back onto Lyell's geology, and he did not label his approach as a theory. However, while he did not consider his approach (which I will call uniformitarian) as a theory or hypothesis, he did see uniformitarianism as offering the very *principles* of the new science of geology. Buckland, like Gould, is right to point out the misreadings around uniformitarianism, as they need to be understood. However, there are further issues for literary and cultural critics concerning what lies behind these misreadings and controversies, issues in and around Lyell's writings and Whewell's coinage of uniformitarianism. It is precisely because the term was coined by Whewell in direct response to Lyell that it is important for understanding the ideological problems Lyell attempted to work through in order to establish his uniformitarian 'principles', and for understanding the ideological problems his text went on to raise. As Tennyson's poems are read as rejoinders to both Lyell's and Whewell's rhetoric, uniformitarianism is of vital importance in itself. It is Tennyson's reaction to the material language of Lyell's and Whewell's texts, his perception of their ideological differences and the possibilities that he intuitively sensed in the wider application of uniformitarianism (both as methodology and as law) that is of interest here, and not the degrees to which the two geological positions were practically adopted by contemporary geologists. Tennyson was responding on a textual level to an opposition that Lyell rhetorically creates, and that Whewell condenses into two opposing doctrines. Therefore, his response cannot reflect practices in the geological field, but it can tell us much about Lyell's ambitions for *Principles,*

---

[14] Buckland, *Novel Science* (2013), 25. Buckland offers a major reassessment of nineteenth-century geology and the novel, demonstrating the rich culture from which it emerged. Buckland adds to the re-visioning of the history of geology by suggesting that "Differences between geologists were subtle and complex, determined much less by religious denomination (as is sometimes asserted) than by forms of publication, institutional affiliations, networks of friendships and collaboration, access to travel, and regional differences, and to the exhibitions, museums, lectures, and written works these men produced and consumed" (25).

Whewell's concerns regarding Lyell's geology and the wider cultural implications of these debates. The chapter returns to examine Whewell's response to Lyell. However, before that it explores Lyell's resistance to evolutionary theory and the strategies and arguments he uses against progressive narratives, as these help to explain Lyell's dogged adherence to uniformitarianism.

GEOLOGY'S FINAL LAWS

Lyell's antipathy towards all types of evolutionary theories has been widely noted.[15] In *Principles*, he made a point of refuting the theories of species transmutation put forward at the beginning of the century by the French biologist, Jean-Baptiste Lamarck (1744–1829) in his *Philosophie Zoologique* (1809). Theories of transmutation had to be addressed precisely because Lyell was well aware that his own uniformitarian geology, with its postulation of geological deep time and its emphasis on gradual, hardly perceptible change, could be seen as the key to the unfolding of biological evolution, as Darwin indeed famously realised while reading Lyell. Lyell also understood the implications of evolutionary theories for the status of humanity, as James Secord points out, if "uniform causal laws applied to all other parts of the natural world, why not to humanity itself?"[16] There is no doubt that Lyell found Lamarck's controversial notion of transmutation highly distasteful.[17] Secord suggests that Lyell was "appalled by the religious, political and ethical consequences of evolution".[18] Lyell had read Lamarck's work in 1827 having received a copy

---

[15] For articles specifically on this topic, see, for example, Martin Rudwick, "The Strategy of Lyell's *Principles of Geology*" (1970). Michael Bartholomew, "Lyell and Evolution: An Account of Lyell's Response to the Prospect of an Evolutionary Ancestry for Man" (1972–1973). Michael Bartholomew, "The Non-Progress of Non-Progression: Two Responses to Lyell's Doctrine" (1976).

[16] James Secord, "Introduction," in *Charles Lyell, Principles of Geology* (1997), xxxi.

[17] Lyell's personal religious beliefs figure little in letters and journals written in the 1820s and 1830s, while his letters do indicate his continual struggle against the demands Scriptural orthodoxy imposed on his writing. J.M.I. Klaver suggests that Lyell's "greatest religious influence was probably John Milton, a copy of whose *Paradise Lost* he won in a contest for reciting poetry at school at Midhurst". Lyell's interest in Milton stems probably from its "dramatic description of life on earth, which could not but appeal to the imaginative mind of a scientist". See, *Geology and Religious Sentiment* (1997), 15.

[18] Secord, *Charles Lyell* (1997), xxxiii.

from Gideon Mantell, but he professed to have felt "none of the *odium theologicum* which some modern writers in this country have visited him with".[19] Nevertheless, he understood transmutation to be a dangerous idea with social and political implications as Lamarck's theory had already been "vilified as the work of an atheist and ... linked to revolution and immorality", the "name Lamarck, French and atheist, [becoming] equated with blasphemy and godlessness".[20] Lyell dedicated the first two chapters of the second volume of *Principles* to refuting Lamarck, after which he went on to expound his own ideas on the viability of species boundaries and the laws regulating the geographical distribution of species. Lamarck was an easy target; he had been humiliated in his own lifetime by the French scientific fraternity. Now dead, he could conveniently stand in for all those evolutionary and directionalist theories that Lyell sought to debunk. Yet instead of diverting attention away from evolutionary theory, *Principles* took Lamarck, by default, to a readership he certainly would not have otherwise reached.[21]

Lyell's rejection of species mutability seems to be part of his own conservative concerns to maintain the boundaries that protect not only man's superiority in creation but also the existing hierarchy of social classes. Adrian Desmond shrewdly observes that "ruling gentlemen had of course a rather literal insight into man's exalted birth; after all, history for them was a continuity of noble descent, not a chronicle of working-class emancipation".[22] Transmutation implied that there was an insidious mobility and disorder in the natural world. Hugh Miller's *Old Red Sandstone* (1841) made the connection more directly, as suggested in Chap. 1, by admonishing Chartist demands for suffrage before turning to Lamarck with the usual Francophobic rebuke of transmutation. Such connections in a geological text attest to the subversive power of evolutionary theory and its association with social discord and anarchy.

Michael Bartholomew rightly attributes "Lyell's distaste for anthropoid origins" to "'psychological', or perhaps aesthetic" reasons that do

---

[19] Lyell to Mantell, March 2 (1827) *Life Letters and Journals of Sir Charles Lyell, Bart*, 2 vols., Katherine Lyell ed. (1881), I, 168.

[20] Klaver, *Geology* (1997), 51–2.

[21] Secord suggests Lyell's criticism of Lamarck in the *Principles* volume two "made evolutionary theory accessible throughout the English-speaking world, beyond a narrow circle of naturalists, medical lecturers and political radicals" *Charles Lyell* (1997), xxx.

[22] Adrian Desmond, *The Politics of Evolution* (1992), 329.

"not stem from a specific theological dogma".[23] However, Lyell's concern with the implications of progressive theories were also to do with his self-perception and his ambitions for *Principles*. Lyell strenuously denied the possibility of mutable species boundaries: If Lamarck believes, he writes,

> that reason itself, may have been gradually developed from some of the simplest states of existence, – that all animals, that man himself, and the irrational beings, may have had one common origin; that all may be parts of one continuous and progressive scheme of development from the most imperfect to the most complex; in fine, he renounces his belief in the high genealogy of his species, and looks forward, as if in compensation, to the future perfectibility of man in his physical, intellectual, and moral attributes. (*PG*, II, 20–1)

Lyell was reluctant to accept the idea that humanity was unable to achieve perfection in the present, that the "perfectibility of man in his physical, intellectual, and moral attributes" was merely a future prospect. The notion of purposive, evolutionary progression was untenable because it posited humanity as existing somewhere along an indeterminable continuum. Humanity in the present was unfinished and unperfected, only able to "look forward, as if in compensation, to the future perfectibility". Essentially, such an idea threatened Lyell's own self-image: his perception of himself as at the apex of a non-transmutable order in nature and society. Narratives of progress also threatened to compromise the perfection of uniformitarian laws in the present, laws that Lyell posited as the final laws of geology.

Lyell constructs a long argument that includes evidence from Egyptian mummified cats to contemporary dog breeding to suggest that there has been no movement across species boundaries over time. There is variation, as "We must suppose, that when the Author of Nature creates an animal or plant, all the possible circumstances in which its descendants are destined to live are foreseen, and that an organization is conferred upon it which will enable the species to perpetuate itself and survive under all the varying circumstances to which it must be inevitably exposed" (*PG*, II, 23–4). However, variety merely suggested movement within a finite scope of variability (an argument that Miller later used).

---

[23] Bartholomew, "Lyell and Evolution" (1976), 268.

Lyell's conclusion is that "species have a real existence in nature, and that each was endowed, at the time of its creation, with the attributes and organization by which it is now distinguished" (*PG*, II, 65).

Lyell faced another issue in his anti-directionalist stance which would be more difficult to explain without recourse to narratives of progress. The apparent recent appearance of humans in biological history posed a niggling problem. How could such a remarkable event as the introduction of humanity and human intelligence be explained without removing altogether the premise that the laws in nature are rigidly uniform throughout time? Lyell had to explain what he took to be the exceptional and exalted nature of humankind and its recent appearance, without undermining his uniformitarian premise that physical laws were unwavering. He takes pains to do this in chapter nine, volume one of *Principles*, making a decisive division between the biological/geological worlds and the human moral and intellectual world—what can be termed for convenience, his 'strategy of division'. With the arrival of humanity, Lyell explained, "so new and extraordinary a circumstance arose, as the union, for the first time, of moral and intellectual faculties" (*PG*, I, 156). According to Lyell's division, it was entirely possible for the physical constitution to deteriorate, while humanity's rationality increased. In turn, Lyell argued that the introduction of humans as physical beings on the earth had had no special influence on the rest of organic creation of which humanity, taken in its physical condition, is a part. Lyell writes:

> If then the physical organization of man may remain stationary, or even become deteriorated, while the race makes the greatest progress to higher rank and power in the scale of rational being, the animal creation also may be supposed to have made no progress by the addition to it of the human species, regarded merely as a part of the organic world. (*PG*, I, 155)

Thus, the "introduction at a certain period of our race upon the earth, raises no presumption whatever that each former exertion of creative power was characterized by the successive development of *irrational* animals of higher orders". Even if "the animal nature of man" is "considered apart from the intellectual", 'man' is still of "higher dignity than any other species" as the comparison between humans and animals, Lyell argues, is "between things so dissimilar, that when we attempt to draw such inferences, we strain analogy beyond all reasonable bounds" (*PG*, I,

155–6). Humanity's creation represented a special event, one outside the laws that govern all other natural events. He argues:

> If an intelligent being ... after observing the order of events for an indefinite series of ages had witnessed at last so wonderful an innovation as this, to what extent would his belief in the regularity of the system be weakened?—would he cease to assume that there was permanency in the laws of nature?—would he no longer be guided in his speculations by the strictest rules of induction? (*PG*, I, 163–4)

Lyell's retort is to suggest that the logical conclusion of such reasoning would be to identify in what ways humanity is in fact different from other creations. The intelligent being

> would undoubtedly be checked by witnessing this new and unexpected event, and would form a more just estimate of the limited range of his own knowledge, and the unbounded extent of the scheme of the universe. But he would soon perceive that no one of the fixed and constant laws of the animate or inanimate world was subverted by human agency, and that the modifications produced were on the occurrence of new and extraordinary circumstances, and those not of a *physical*, but a *moral* nature. (*PG*, I, 164)

Clearly, as Roy Porter suggests, "for Lyell there was a deep ontological dualism between the history of the earth and the history of man, because the one is the natural and the other the moral and rational realm".[24] Humanity, he believed, was exceptional because of its moral sense and its reasoning capacity, which does not come within the compass of the laws of nature.

Another difficult hypothesis Lyell had to deal with in order to discount progressive theories was the 'embryonic' or 'recapitulation' theory of German anatomist and physiologist Friedrich Tiedemann, which he addresses in volume two, chapter four of *Principles*. Tiedemann had discovered that

> the brain of the fœtus, in the highest class of vertebrated animals, assumes, in succession, the various forms which belong to fishes, reptiles, and birds, before it acquires those additions and modifications which are peculiar to

---

[24] Porter, "Charles Lyell", 97.

the mammiferous tribe. So that in the passage from the embryo to the perfect mammifer, there is a typical representation, as it were, of all those transformations which the primitive species are supposed to have undergone, during a long series of generations, between the present period and the remotest geological era. (*PG*, II, 63)

Lyell concedes to the persuasiveness of such stages of foetal development, and to their analogous compatibility with theories of transmutation. However, he argues for non-progression and the fixity of species by inferring that these stages suggest not progress but rather limited movement within a finite spectrum of potentiality:

If we could develop the different parts of the brain of the inferior classes, we should make in succession a reptile out of a fish, a bird out of a reptile, and a mammiferous quadruped out of a bird. If, on the contrary, we could starve this organ in mammalian, we might reduce it successively to the condition of the brain of the three inferior classes. Nature often presents us with this last phenomenon in monsters, but never exhibits the first. (*PG*, II, 63)

Embryonic development "lend[s] no support whatever to the notion of a gradual transmutation of one species into another ... On the contrary, were it not for the sterility imposed on monsters, as well as hybrids in general [... embryonic development] would be in favour of the successive degeneracy, rather than the perfectibility, in the course of ages, of certain classes of organic being" (*PG*, II, 64). Embryonic theory is thus made not only to undermine theories of transmutation but also to support Lyell's fundamental belief that there can be no progress in organic nature, nor any in the hierarchy of men—man's highest physical potential is already available in the present, as is his potential to degenerate, but there can be no progression beyond the full potential of human physical perfection. Lyell used the embryonic hypothesis (which Chambers would later use in *Vestiges of the History of Creation* (1844) to support progression in both biological and intellectual terms) to argue that there is no progress, only change contained by the bounds of a finite system, and his argument is consistent with his wider uniformitarian view of non-progressive geological change. The serial layering of distinct strata with the passing of each separate geological epoch may seem progressive but, in uniformitarian terms, it is merely descriptive of the continual re-cycling of what already exists in the earth's system through the unending exertion of uniformitarian geological laws. Not only does this non-progression system help to steer thinking away from evolutionary theories, it also serves Lyell's wider perceptions of the social order and his own position in it. His non-progressive rhetoric affirms and helps to maintain the status quo as if the past is no different from the present, the future, in turn, will be much like the present and no different from the past.

LYELL'S "HISTORY OF GEOLOGY"

Lyell's refutation of Lamarck, his arguments against successive and progressive creations and against embryonic theory were not only the result of his distaste for what evolution meant for 'man's' "high genealogy", they were also part of his wider strategy and his ambitions for *Principles*. Encoded in the rhetoric of his 'history of geology', which opens the first volume of *Principles*, was a covert agenda aimed at establishing uniformitarian thinking as a new, anti-catastrophist and rational mode of thinking. Lyell capitalises on catastrophism's association with Mosaic history in order to discredit any recall to catastrophic events in the geological history of the earth, promoting instead an insistence on uniformitarian laws of nature. Although Lyell is careful not to make a direct connection, catastrophism becomes associated with the attempt to keep geology in harmony with religious doctrines at the expense of the observable facts. Roy Porter suggests that Lyell saw geology as torn between the "polar opposites of Hellenistic uniformitarianism, which was the only true scientific method, and Hebraic catastrophism, which was unscientific, religiously biased [and] geologically fruitless".[25] Aligning catastrophe and irregular causes with irrationality and superstition, Lyell proposed instead a uniformitarian perspective that saw all geological change as operating to fixed and invariable natural laws, this being the only rational way of positing geological change. In this, as Porter says, he "vulgarises the multiplicity of stances between pure uniformitarianism and pure catastrophism".[26] Lyell writes, in an oblique criticism of religious dogma, of the "effects [that] may be produced by great catastrophes ... recurring

---

[25] Roy Porter, "Charles Lyell and the Principles of the History of Geology" (1976), 98.
[26] Ibid., 98.

at distant intervals of time, on the minds of a barbarous and unculti-
vated race" (*PG*, I, 8). Catastrophic geological phenomena play on the
irrational imagination and exert an irresistible power. When "such rare
phenomena is witnessed in the present course of nature, it scarcely ever
fails to excite a suspicion of the preternatural in those minds which are
not firmly convinced of the uniform agency of secondary causes;—as
if the death of some individual in whose fate they are interested, hap-
pens to be accompanied by the appearance of a luminous meteor, or a
comet, or the shock of an earthquake" (*PG*, 1, 98). The implication is
that catastrophe goes hand-in-hand with irrationality. Catastrophe is
measured in terms of the individual's experience rather than in the wider
terms of the history of the planet's geology. Catastrophism is irrationally
anthropocentric and suggestive of a Romantic self-indulgence that projects its imagination on to nature. Thus, enlightened detachment is pit-
ted against a Romantic propensity to validate subjective experience. To
see catastrophes as random, irregular and outside the fixed laws of geological process, Lyell suggests, is intrinsically unscientific. Alternatively,
uniformitarian thinking is scientifically sound thinking because it resolves
catastrophe, via deep time, into a stable vision of an earth responding to
measured, regular laws. "By degrees" Lyell writes, "many of the enigmas of the moral and physical world are explained, and, instead of being
due to extrinsic and irregular causes, they are found to depend on fixed
and invariable causes" (*PG*, I, 76). Fixed laws will eventually explain all
geological phenomena and the smooth path of uniformitarian thinking
will eventually eradicate the Romantic desire to subsume nature into the
self, and the primitive urge to reconcile geological change with religious
doctrines. Uniformitarian thinking ends the strangely retrogressive pull
of the human psyche, perfecting not merely the study of geology but the
human mind itself, and, in the process, ending the need for progress,
evolutionary or otherwise.

The "history" works hard to associate the geological idea of catastrophism with an archaic, anthropocentric and irrational view of the
earth. For Lyell, catastrophism is not merely an erroneous theoretical
framework for interpreting geological phenomena—it is a psychological condition describing an archaic mentality. Catastrophism is linked
to a "half-civilized", "rude state of society"; it is a remnant of the past
that lives on in the layers of the human psyche. Just as the mind must
be purged of erroneous beliefs, so the aim of the "history" is to recount
the past in order to discount it—to bring the archaic into consciousness

to be thereafter acknowledged as erroneous by the reasoning mind and once and for all discarded. Lyell brings the same logic to his thinking in the "history" as he does to all progressive theory. Thus, he suggests the "superstitions of a savage tribe are transmitted through all the progressive stages of society, till they exert a powerful influence on the mind of the philosopher" (*PG*, I, 8). As inferior forms indwell in the embryo, the "puerile conceits and monstrous absurdities" (*PG*, I, 6), those oldest relics of the human psyche, remain to exert a retarding influence on the modern mind. The human mind, like the embryo, can fulfil its potential and ascend to the perfection of human intellectual achievement, or it can stagnate or even regress back into its lowest state. Thus, the human psyche, which contains all human potential, also contains the primitive mind which represents its lowest mental capability, just as embryonic theory suggests a potential variation albeit within the bounds of individual species. Improvement is not indicative of progress but suggests possibility within a finite scale—a spectrum in which perfection already exists, albeit in a state only attainable by a superior species and only the superior individuals of that species. Lyell ostensibly solved the problem of evolutionary progress and its inference of a "future perfectibility" that is always unattainable, and by refusing a progressive narrative of nature he satisfies his antipathy towards evolutionary theories, his conservative propensity towards upholding the social order, his ambitions for *Principles* as ending the finite history of geology and his wider vision of fixed uniform laws.

Lyell, however, was not the first to make a case for 'uniformitarian thinking' as a paradigm for rationality. He follows James Hutton, who, in his *Theory of the Earth* (1797), also theorised from a uniformitarian premise. Hutton's aim was to propose, in Newtonian manner, a set of natural, regular and immutable laws by which all geological change is governed. He suggested such uniform laws are to be found in what he argued were the consistent processes by which the earth is continually and simultaneously eroding and decaying, reforming and renewing itself—processes which are always in perfect harmony.[27] In Enlightenment fashion, Hutton also insinuated, as Lyell would later, that there was an inherent irrationality in any explanation of geological phenomena that suggested nature's processes were random and

---

[27] James Hutton, *The Theory of the Earth with Proofs and Illustrations*, 3 vols. (1795). Facsimile edition (1959).

unsystematic, rather than uniform. Where catastrophe is understood as an aberration of normative patterns in nature, in Hutton's and Lyell's thinking, erroneous beliefs and superstitions are rife. Uniformity, however, dispels superstition and normalises such phenomena, working even the most aggressive and destructive events into a wider harmonious system of ruin and renewal. Uniformitarian thinking rejected the view of geological change as intermittent, haphazard and uncontrollable, re-visioning catastrophe through the lens of deep time. Uniformitarianism was an enlightened, composed and gentlemanly way of thinking exercised by the rational mind that is capable of managing the geological sublime rather than being overwhelmed by it. Thus uniformitarianism released the rational thinker from the tyranny of what appeared to be the unpredictable and irrational operations of capricious and destructive forces, offering in its place an ordered and balanced world that moved through time abiding by regular and immutable laws.[28]

Perhaps the best example of a politically orientated geological uniformitarianism comes from the writings of George Hoggart Toulmin (1754–1817). Regarded as having little influence on subsequent uniformitarians, Toulmin's geologic system was nevertheless closely aligned with Hutton's.[29] Whether Lyell read Toulmin is not known, but it can be assumed that even if he had he would not have acknowledged him, as Toulmin's uniformitarianism was mobilised in the service of his radical politics and atheism. As Roy Porter asserts, Toulmin's theory of the earth is an "index of the intimate ideological interpretation of political commitment and geological science at the end of the Enlightenment and the beginning of the Romantic Age".[30] Uniformitarianism was for Toulmin a superior and rational way of thinking and an alternative to ideas of Mosaic catastrophe. It challenged catastrophist models that appeared to suggest that the earth was fixed and immutable, responding only to punitive divine intervention manifested in unpredictable and

---

[28] For a reading of Percy Bysshe Shelley's poetic engagement with Hutton's uniformitarian geology, see Geric, "Shelley's 'cancelled cycles': Huttonian Geomorphology and Catastrophe in *Prometheus Unbound*" (2013).

[29] Porter suggests "It is well known that Toulmin's ideas bear close resemblance, in their uniformitarianism, to those of Hutton and his followers, and scholarly debate has raged as to whether Toulmin plagiarized from Hutton, or vice versa." "Philosophy and Politics of a Geologist: G.H. Toulmin (1754–1817)" (1978), 437n.

[30] Ibid., 436

indiscriminate destruction. Toulmin set out unambiguously to "shake the fixed and the malevolent prejudices of mankind; to assuage the remaining turbulence of ignorance and error; and thus to smooth the way to ... civilization and refinement".[31] Lyell takes up this view of uniformitarianism as a superior, refined way of thinking that better befits a gentlemanly mentality unimpressed by the vulgar sublime. Thus, while Toulmin's politics differed from Hutton's and Lyell's in its radicalism, all three expounded a dichotomy in which uniformitarianism stood as the superior, civilised and sophisticated alternative to catastrophism's retarding irrationality.

There were, however, problems with Hutton's uniformitarian patterns of change that Lyell would inherit. Hutton's geomorphology described the earth as ostensibly permanently imprisoned within non-progressive change. Uniformitarian processes worked incessantly in the present to keep the system of the earth in a state of harmony. The seemingly endless re-cycling of the earth's material substance infected Hutton's writing, which seemed, like his theory, repetitive and non-progressive. Lyell, so much more sophisticated in his rhetoric than Hutton, understood the pitfalls of postulating a non-progressive system in which the narrative can so easily be lost and, in consequence, human interest too. Lyell knew that with such a loss of linearity he risked alienating his readership, and his "history" was designed in part to counter the non-progressive nature of the uniformitarian vision—to give a narrative to uniformitarianism's non-progressive story. Depicting geological change as non-progressive, his aim was nevertheless to show a progression in geological science by recounting the history of human thought on geology. At the same time, Lyell sought to avoid the implication that his own theories were merely conclusions drawn from previous geological ideas, and that they in turn would be superseded in time by newer, more advanced theories. The "history" begins with the remote past of human thinking and ends with *Principles* itself. As Porter says, Lyell "thought himself offering—for the first time—the very principles of reasoning in the science", principles that would put an end to the history of the theory of geology.[32] Lyell achieves this effect through a rhetorical strategy that utilises the structure

---

[31] G.H. Toulmin, *The Eternity of the Universe* (1789) from the final page of the unnumbered Introduction.

[32] Roy Porter, *Charles Lyell* (1976), 92.

of the theory of uniformity itself. He does not add to the history of geology; instead, he effects a kind of end to that history by leading his reader to the end of a linear path (his "history") where they encounter uniformitarian principles as the final word on geology. Applying the logic that he uses for all progressive narratives—evolution, embryonic theory, the human psyche—Lyell places his history within a finite spectrum of geological thought and discovery, and he takes his readers from the lowest to the highest point of perfection—to uniformitarianism itself. The geological history does exactly what Lyell wants it to do; it offers readers a narrative, but it ends that narrative with *Principles*, creating a time-arresting moment—a uniformitarian moment—in which everything knowable is known in the present via the uniformitarian principles Lyell sets out. Thus, Lyell provides the final laws in geological science, laws both advocating non-progression and unable to be progressed.

After the linear narrative of the 'history of geology', volume one falls into a different pattern, levelling out into a concentrated examination of present causes. Having conquered the past and shown it to be inadequate for understanding of the earth's structure and processes of geological change, Lyell begins his steady and relentless accumulation of observable evidence in the present. His aim was to demonstrate, with a meticulous attention to material evidence drawn from his own field work and that of other geologists globally, that accumulated actions over the span of deep geological time can produce the most profound geological change. The 'history of geology' was evoked and expelled in order to make way for Lyell's uniformitarian argument that it is only from the observation of evidence in the present that a true understanding of geomorphology can be acquired. To establish this, the rest of volume one offers layer upon layer of evidence drawn from the examination of the effects of aqueous and igneous agents.

Crucially, another of Lyell's aims was to argue that aqueous and igneous forces operate against each other to effect change that, paradoxically, keeps the system of the earth in perfect balance. Thus, igneous agents—forces such as uplift, earthquake, volcano—Lyell argues, perform a vital service in keeping in check the power of aqueous agents—such as glacial erosion. The earth is in a continual state of flux in which the loss of land caused by erosion is balanced out by the effects of igneous agents that work to uplift the land, and it is this balance that keeps the planet habitable. In such a view, the forces of catastrophe are re-visioned as stabilising, constructive forces essential to the overall health of the system. For example, in discussing igneous agents, Lyell concludes:

> it appears ... respecting the agency of subterranean movements, that the constant repair of the dry land, and the subserviency of our planet to the support of terrestrial as well as aquatic species, are secured by the elevating and depressing power of earthquakes. This cause, so often the source of death and terror to the inhabitants of the globe, which visits, in succession, every zone, and fills the earth with monuments of ruin and disorder, is nevertheless, a conservative principle in the highest degree, and, above all others, essential to the stability of the system. (*PG*, I, 479)

Having set out his empirical credentials in volume one, in the next volume Lyell moved on to the tackle some of the grittiest problems that fed into geological study: evolutionary theory, the creation and extinction of species, the geographical distribution of species, his controversial theory of climate change—before ending with an extended examination of processes of fossilisation and the "imbedding of organic remains". In volume three, Lyell reassessed his overall arguments before applying the evidence of volumes one and two to the "decipherment of the past history of the earth".[33] *Principles* is such a rich text that this chapter cannot do it justice. However, an important aspect of *Principles* for Tennyson's work was the subtle way that Lyell converted readers to uniformitarian thinking by erasing the evidence of the past—the history of geology—with evidence found in the present—converting them, in other words, to see the present as the key to the past.

Lyell's uniformitarianism was not only a way of seeing geological change; it represented a modern and enlightened mentality, of which Lyell's own reasoning was the consummate example. Overthrowing catastrophist thinking and evolutionary theories, his aim was to imply that no further progress could be made in either the principles of geology or in the quality of the geologist's mind. Lyell's reasoning recalls Edmund Burke's reactionary rebuttal of revolutionary notions of political progress. Where Burke insisted that "no discoveries are to be made in morality, nor many in the great principles of government, nor in the ideas of liberty", Lyell's *Principles* establishes a style of thinking from

---

[33] Rudwick, Introduction to *Principles* (1990), vol. I, xxxviii.

which no progression is necessary.[34] Lyell discovers the uniformitarian *principles* that were already there, and that, now discovered, are permanent laws, perfectly comprehensible to those whose tone of mind is also uniformitarian.

In *Principles*, Lyell was working towards nothing less than a paradigmatic reappraisal of the nature of knowledge, one that can only be achieved by a mind released from the linearity that connects it with the archaic psyche—in other words, a rational, stable, uniformitarian mind, freed from the retarding pull of the past, a mind able to see with clearsighted vision the operations occurring in the present. Lyell's approach to non-progressive and progressive models of change are complex. Uniformitarianism describes an apparently non-progressive system, and at the same time, uniformitarian thinking is not progressive thinking; it is thinking that has progressed to its full capacity within the finite scale of human potential—a scale represented both in the variability of species within set bounds and the forms suggested by embryonic theory. It is this denial of unlimited progress, performed rhetorically partly in order to avoid any postulation of evolutionary theory and partly to establish himself and his principles as unable to be progressed, which makes Lyell's thinking, his science and his text intrinsically and problematically anti-progressive. Lyell puts himself at the apex of the development of human thought. He explodes an older world of catastrophe, re-forming it ostensibly into a stable, predictable uniformitarian world, where the past is made subservient to the present and where it is ordered and made knowable by reference to present processes. In *Principles*, the forces of geological change are ordered into a marvellous symmetry in which seemingly antagonistic aqueous and igneous agents work in unison, and in unending time, levelling and subordinating partisan systems and doctrines which are brought under the triumphant and unassailable banner of uniformitarianism. Lyell also recalibrates the mind that apprehends the world with the force of *Principles's* empirical evidence and with his consummate rhetorical skill. He replaces regressive catastrophist thinking (thinking stuck in the archaic, irrational and superstitious strata of the human mind) with the measured, superior reasoning of uniformitarian thinking, and in doing so he takes geology and uniformitarian thinking

---

[34] Edmund Burke, *Reflection on the Revolution in France*, L.G. Mitchell ed. (1993), 86.

into a state of present perfection where there is no progress, where nothing needs progressing, as the principles of geology have been won.

WERNER'S CATASTROPHIC DEFEAT

Lyell summoned up an impressive rhetorical power in his writing, and as Ralph O'Connor has shown, Lyell "used catastrophic imagery to undermine 'catastrophe'".[35] How this actually works can be seen in his treatment of the German mineralogist Abraham Gottlob Werner (1749–1817). Lyell mobilised catastrophe in his offensive against his opponents (and sometimes those who were not necessarily his opponents), and Werner, like Lamarck, was a convenient target, as he was both foreign and dead. Werner had been respected and revered both in Germany and in Britain, with mineralogists and budding geologists flocking to study under him from all over Europe. Werner was, however, on the losing side of the 'Neptunist/Vulcanist' debate, arguing the Neptunist position that the earliest rocks were not of volcanic origin but produced when a universal sea covered the highest mountains.[36] Alexander Ospovat has written about what he sees as Lyell's gross distortion of Werner's theories and methodologies, attempting to rescue Werner from Lyell's criticism and re-establishing his reputation as a man who made a seminal contribution to mineralogy. For example, Ospovat draws attention to what he sees as the "four Werner myths" propagated by Lyell in his "history". They are the "antipathy-to-writing myth, the lack-of-travel myth, the 'onion' myth, and the retrograde-influence-on-geology myth". What Ospovat describes as the "onion" myth is Lyell's interpretation of Werner's theory that strata were uniformly laid down over the entire earth in the same order, which Ospovat suggests is an "oversimplification of Werner's theory of the precipitation and deposition of strata".[37] The "lack-of-travel myth" explained the "onion" myth, as Lyell implies Werner's antipathy to travel and field observation led him to erroneous conclusions.

---

[35] Ralph O'Connor, *The Earth on Show* (2007), 167.

[36] See Martin Rudwick, *Bursting the Limits of Time* (2005), 175.

[37] Alexander Ospovat, "The Distortion of Werner in Lyell's *Principles of Geology*" (1976), 191–2.

Lyell, however, goes much further than merely inferring Werner's methods and theories were suspect. He treats Werner as a 'catastrophist', constructing him as the antithesis of himself as a uniformitarian. Lyell describes "Werner's mind" as "at once imaginative and richly stored with miscellaneous knowledge" (*PG*, I, 55–6). Werner, it appeared, combined two traits that were dangerous to the assertion of scientific fact—a Romantic propensity to cultivate the imagination that would later account for what Lyell saw as Werner's capacity to invent, and a mind occupied with "miscellaneous knowledge" in opposition to one with ordered, detailed and specialised knowledge. Werner's lectures, for example, were "excursive", interlaced with speculations on "the economical uses of minerals, and their application to medicine; the influence of the mineral composition of rocks upon the soil, and of the soil upon the resources, wealth, and civilization of man". Similarly, he would digress into discussions on the effect of the landscape on the "different manners" of its inhabitants, their "wealth and intelligence". Nothing was excluded from Werner's geologic or mineralogical speculation:

> The history even of languages, and the migrations of tribes had, according to him, been determined by the direction of particular strata. The qualities of certain stones used in building would lead him to descant on the architecture of different ages and nations, and the physical geography of a country frequently invited him to treat of military tactics. (*PG*, I, 56)

Lyell sets Werner up as the model of a pre-scientific thinker against his own example of the modern specialist who limits his speculation, and in this he was setting in place the characteristics that would go on to define what it is to be a 'scientist'—the term that Whewell coined in 1833, not long after the publication of *Principles*. According to Lyell, Werner was unscientific because his "theory was opposed … to the doctrine of uniformity in the course of nature". Werner was a catastrophist because he, like catastrophists generally in Lyell's language of opposition, felt at liberty to "introduce, without scruple, many imaginary causes supposed to have once effected great revolutions in the earth, and then to have become extinct" (*PG*, I, 58). Thus, Werner was the Romantic catastrophist whose imaginative productions were attractive and captivating but whose thinking, nevertheless, was the antithesis of rigorous scientific reasoning.

Lyell concedes that Werner displayed a "genius", but not in his contribution to geology or even stratigraphy or mineralogy. Rather, what distinguished him in the history of geology was the "charm of his manners", his "eloquence", and his ability to "kindle enthusiasm in the minds of all his pupils", to "persuade others to believe" and to "inspire all his scholars with a most implicit faith in his doctrines" (*PG*, I, 56–7). Werner inspired *faith* and *belief*, but failed, by inference, to encourage his students to develop their own observational or reasoning faculties. It was, according to Lyell, "a ruling object of ambition in the minds of his [Werner's] pupils to confirm the generalizations of their great master, and to discover in the most distant parts of the globe his 'universal formations'" (*PG*, I, 57). Thus, the "ruling object" that the "great master" installed in his disciples was merely one that furthered Werner's own ambitions and endorsed his opinions, which were, nonetheless, no more than "generalizations". Lyell's language was also redolent with religiosity: Werner sent out his followers, like disciples, to the "most distant parts of the globe" to spread the word of the great master's doctrines. They, in turn, exhibited the "fullness of their faith" by supporting erroneous theories: they were "blinded by their veneration for the great teacher" and displayed an evangelical "impatience of opposition" (*PG*, I, 60). Werner was the high priest of the cult of Neptunism; he was the "great oracle of geology" who inspired blind faith in his followers so their one "ruling object" was the furtherance of what Lyell paints as Werner's monomaniacal ambitions (*PG*, I, 57). Thus, Werner's followers, in the mode of an Inquisition, "tortured the phenomena of distant countries, and even of another hemisphere, into conformity with his theoretical standard". A dogmatic, almost tyrannical personality emerges from Lyell's description. Werner was the Gothic villain; the dangerous, charismatic Romantic who sent out his agents to do his bidding. His theories were absolute, the "supreme authority usurped by him over the opinions of his contemporaries, was eventually prejudicial to the progress of the science, so much so, as to greatly counterbalance the advantages which it derived from his exertions". Rather than advancing science, Werner's influence retards science, contaminating it with a quasi-religious fanaticism (*PG*, I, 57, 56).

Tellingly, however, it was Werner the man who was implicitly criticised in Lyell's "history", as Lyell implies a relationship between Werner's theories and his character. We are told that although the "natural modesty of [Werner's] disposition was excessive, approaching even to timidity, he indulged in the most bold and sweeping generalizations" (*PG*, I, 56). This, then, was a man whose science was coloured by his character, but subtler is the inference that Werner was at once timid and bold, indicating inconsistency and instability. Werner was a man of contradictions, a man with an uneven temperament, who was not only a catastrophist in his geological theories but was catastrophic by nature in the excessive contradictions of his personality. Werner's thinking was catastrophist thinking, and this, in turn, had catastrophic consequences for geology. Thus, in revolutionary and catastrophic style Werner "overturned the true theory", and replaced it with "one of the most unphilosophical ever advanced in any science" (*PG*, I, 59). Werner's erroneous adherence to Neptunism and to catastrophism (according to Lyell) retarded the progress of uniformitarian theories, and "the *tide* of prejudice *ran* so *strong*", against those who held to the opinion of uniformity (Hutton and Playfair) that these more reasonable thinkers were "*carried* far away into the chaotic fluid, and other cosmological inventions of Werner" (*PG*, I, 69). While the "fictions of the Saxon Professor" swallowed up the uniformitarians in an apocalyptic deluge of error and imprecision, Lyell appropriates the language of catastrophe in order to discredit Werner, his character, his science and all those who follow his geological fictions.[38] Lyell's treatment of Werner is just one example among many of the great investment Lyell placed in his rhetorical skills. Lyell's "history" was a masterstroke of propaganda that caused various misunderstandings in the history of geology for over a century and that associated catastrophism with unscientific ways of thinking long into the twentieth century. In spite of Lyell's sometimes clumsy distortions (such as his theory of climate change as a way to account for the appearance and extinction of different species), he managed for enough time to claim for himself the reasoned ground of science in geology via his characterisation of uniformitarianism, largely because his rhetoric has been left unexamined.

---

[38]For an in-depth reading of Lyell's literary style, see Ralph O'Connor, *The Earth on Show* (2007), 163–87.

"Uniformitarian Arguments Are Negative Only"

William Whewell was perhaps Lyell's most able critic. He read *Principles* and in response coined the terms 'uniformitarianism' and 'catastrophism' in his monumental three-volume *History of the Inductive Sciences* (1837). Behind Whewell's coinages was a complex understanding of Lyell's ambitious agenda and a shrewd awareness of the implications of Lyell's uniformitarianism. Rather than dismantling Lyell's oppositions, Whewell turns them to his advantage, and this is particularly evident in his *Indications of the Creator* (1845), which was largely made up of salient extracts from his own "History of Geology" in his *History of the Inductive Sciences*. *Indications* was published in direct response to the success of Robert Chambers's *Vestiges of the History of Creation* (1844), and thus the text has two contrasting aims: to argue against Lyell's insistence on uniformitarian fixed laws and to counter the developmental thrust of *Vestiges*. Whewell writes of the "*Doctrine of Uniformity*", which he sums up as an approach that sees "the universal action of causes which are close at hand to us, operating through time and space without variation or decay".[39] Such an approach asks the geologist to theorise from the premise that there has been no variance in the type and rate of geological change from that which is presently observed. The "*Doctrine of Catastrophes*", alternatively, is defined as "see[ing] in the present condition of things evidences of *catastrophes*; changes of a more sweeping kind, and produced by more powerful agencies than those which occur in recent times. Geologists who held [this] opinion, maintained that forces which have elevated the Alps or the Andes to their present height could not have been any forces which are now in action" (*Indications*, 147–8). Under the telling heading "Uniformitarian Arguments Are Negative Only", Whewell writes:

> There is an opposite tendency in the mode of maintaining the catastrophist and the uniformitarian opinions, which depends upon their fundamental principles, and shows itself in all the controversies between them. The Catastrophist is affirmative, the Uniformitarian is negative in his assertions: the former is constantly attempting to construct a theory; the latter delights in demolishing all theories. The one is constantly bringing fresh

---

[39] William Whewell, *Indications of the Creator: Extracts from The History of the Inductive Sciences* (1845), 147–8.

evidence of some great past event, or series of events, of a striking and definite kind; his antagonist is at every step explaining away the evidence, and
showing that it proves nothing.[40]

Whewell's affirmative/negative dichotomy posits Lyell as he who
"delights in demolishing all theories". Thus, Lyell is cast as the negative
protagonist in Whewell's history of geology. What had been for Lyell
his modern, empirical, rational stance—his unassailable 'uniformitarian thinking'—becomes, in Whewell's rhetoric, the obstinate posturing
of antagonistic negativity. Like a uniformitarian steamroller levelling the
nuanced undulations of the past, for Whewell, Lyell's *Principles* unscien-
tifically insisted that the 'present is the key to the past'—that, geologically speaking and in uniformitarian terms, the past is the same as the
present and the future will also be the same as the present as the agents
of change are uniform. Where catastrophists supply "fresh evidence",
Lyell, it is implied, "explain[s] away" that "evidence", countering it
not with the power of more persuasive evidence but with something
less empirical and far more slippery—the persuasive force of language.
Whewell thus responds to the impressive rhetoric of *Principles* with his
own. But his challenge to Lyell goes much further, offering an excel-
lent opportunity to understand the wider implications of *Principles* and
exactly what was at stake in the language of geology.

Whewell was an astute reader of Lyell. Fully endorsing Lyell's insistence on observation of the present, he, nevertheless, criticises his refusal
to acknowledge observable evidence that seemed to suggest different
types or rates of change in the past.[41] His criticism was not only based
on his recognition that an insistence on the reality of uniformitarianism
as an unwavering physical law (not merely as a methodological premise) was unscientific but also, and more fundamentally, on uniformitarianism's incompatibility with teleological epistemologies, and Whewell's
construction of affirmative/negative characters for the catastrophist and
the uniformitarian need to be seen in this light. Where Lyell appeared
(particularly to early historians of geology) to theorise from observed
evidence, attempting to dissociate geology from Mosaic history in order

---

[40]William Whewell, *Indications* (1845), 155.

[41] Interestingly, Whewell argued for teleological development using the 'nebular hypothesis' and the evidence of chemistry.

to define it as a distinct scientific discipline, Whewell was attempting to subsume geology into his wider conception of the unity of knowledge. Whewell ably showed how Lyell's insistence on uniformity as a substantive reality posited him at a less than scientific extreme, as demonstrated by his succinct reasoning in his 1837 reply to *Principles*:

> It must be granted at once, to the advocates of this geological uniformity, that we are not arbitrarily to assume the existence of catastrophes. The degree of uniformity and continuity with which terremotive forces have acted, must be collected, not from any gratuitous hypothesis, but from the facts of the case. We must suppose the causes which have produced geological phenomena, to have been as similar to existing causes, and as dissimilar, as the effects teach us. (*HIS*, III, 513)

Whewell went on to qualify: "when Mr. Lyell goes further, and considers it a merit in a course of geological speculation that it *rejects* any difference between the intensity of existing and of past causes, we conceive that he errs no less than those whom he censures". Uniformitarianism could not be assumed to be the ruling law of all geological change, and where empirical facts and observable data suggest a degree or rate of change not observed in the present, that evidence must be considered rather than dismissed for the sake of uniformity. Lyell must have seen the logic of this, which is why he was careful not to frame his 'principles' overtly as a 'doctrine of uniformity', knowing it would be difficult to defend such a position. Whewell's coinage reveals Lyell's difficulty, and demonstrates uniformity's resultant negativity. Lyell was compromised, as to directly place *Principles* under the banner of outright uniformitarianism would have revealed weaknesses in his geology. However, any evidence for different types of geological change or more intense geological activity in the past would ruin the wonderful symmetry of his uniform geological world and his claims for uniformitarian thinking. Lyell had pinned everything on the unwavering laws of uniformity, as for Lyell *Principles* would not be *the* principles if he had only used uniformitarianism as a methodology.

Whewell, however, had his own biases. His mentality was fundamentally different from Lyell's, and the two men make an interesting and important comparison. Where Whewell's polymathic stance was part and parcel of his adherence to the unity of knowledge, Lyell rejected the polymathic model; his "imagined ideal", as Martin Rudwick suggests, "was not a polymath, but rather a savant", a man of science who "worked within a network of others who he 'pumped' for diverse forms of specialized knowledge".[42] Against Lyell's Enlightenment stance, Whewell offered a Romantic sense of the cohesion of knowledge that worked on a teleological premise. He saw patterns running through both physical nature and the human intellectual realms, as in his reckoning, all creation was directional, moving from first causes towards a divine and prefigured final cause. Whewell took exception, for example, to what he took to be the cyclical repetitiveness of Lyell's system, because such cycles effectively abolished history, purpose and design. Arguing against Lyell, Whewell asks:

> Is it not clear ... that history does not exhibit a series of cycles, the aggregate of which may be represented as a uniform state, without indication of origin or termination? Does it not rather seem evident that, in reality, the whole course of the world, from the earliest to the present times, is but *one* cycle, yet unfinished;—offering, indeed, no clear evidence of the mode of its beginning; but still less entitling us to consider it as a repetition or series of repetitions of what had gone before? (*HIS*, III, 517)

The vision of a geological world moving through cycles with no direction was untenable for Whewell, whose mentality was geared towards design. His aim was to reinstate a directional and teleological sense of purpose and to demonstrate how such purpose was knowable in the human mind. To this general end, Whewell coins the term 'palætiology' to describe those disciplines (of which geology, Whewell explains, is the best representative) that attempt "to ascend to a past state of things, by the aid of the evidence of the present" (*HIS*, III, 398). The palætiological sciences unite disciplines from both the natural and the moral spheres: where the palætiological sciences are concerned, "speculation is not confined", Whewell writes, "to the world of inert matter; we have examples of them in inquiries concerning the monuments of the art and labour of distant ages; in examinations into the origin and early progress of states and cities, customs, and languages" (*HIS*, III, 397). He goes on to assert:

---

[42] Martin Rudwick, *Worlds Before Adam* (2008), 300.

The history of the earth and the history of its inhabitants ... are governed by the same principles. Thus the portions of knowledge which seek to travel back towards the origin, whether of inert things or the works of man, resemble each other. Both treat of events as connected by a thread of time and causation. In both we endeavour to learn accurately what the present is, and hence what the past has been. Both are *historical* sciences in the same sense. (*HIS*, III, 402)

While uniformitarianism should not be assumed to offer the unwavering physical laws of geological agency, as a methodology it works for Whewell in important ways. Crucially, the palætiological sciences are all united by the uniformitarian methodology, as such a methodology reveals a commonality between all the historical disciplines. The uniformitarian method indicates an underlying intelligence by offering observers not the physical laws of nature but the intellectual means of apprehending unity in nature, a capacity Whewell saw as immanent in the human psyche.

Key to this way of thinking was the fact that while the uniformitarian methodology tells of "the present state of things" as a means of "throwing light upon the causes of past changes", it cannot throw light on the nature of first causes (*HIS*, III, 399). Thus, the palætiological disciplines all share the fact that their origins cannot be accounted for through uniformitarian reasoning. Where Lyell had suggested that speculations on first causes were not the remit of uniformitarian geology, Whewell used this apparent limitation of the uniformitarian methodology as evidence of the common divine origin of both physical and metaphysical dimensions. He makes, in effect, a virtue of the negativity of the uniformitarian doctrine by suggesting that its inability to throw light on first causes proves that such causes are beyond the scope of material science:

All palætiological sciences, all speculation which attempts to ascend from the present to the remote past, by the chain of causation, do also, by an inevitable consequence, urge us to look for the beginning of the state of things which we thus contemplate; but in none of these cases have men been able, by the aid of science, to arrive at a beginning which is homogeneous with the known course of events. The first origin of language, of civilization, of law and government, cannot be clearly made out by reasoning and research; just as little, we may expect, will our knowledge of the origin of the existing and extinct species of plants and animals, be the result of physiological and geological investigation. (*HIS*, III, 483)

The palætiological sciences converge in the remote past, and what they have in common is that their origins are beyond scientific observation or speculation. Therefore, Whewell writes, respecting origins "we may be unable to arrive at a consistent and definite belief, without having resource to other grounds of truth, as well as to historical research and scientific reasoning", 'other grounds of truth' being, for Whewell, theological and metaphysical (*HIS*, III, 484).

Fundamental to Whewell's development of the palætiological category of sciences is his aim to discover an underlying link between all disciplines, an objective conceptualised in Whewell's 'consilience of inductions'. Consilience, another of his coinages, is a product of his teleological premise. It describes a unity of knowledge based on the concept that there is a common foundation to all branches of knowledge, the palætiological category of sciences being an example of Whewell's sense of the approximation of all knowledge. They offer evidence of the existence of what Whewell called "general truths"—those being truths that can be known *a priori*. Empirical facts and *a priori* ideas meet in the palætiological disciplines, which are in turn all linked by the negative evidence of the uniformitarian methodology, as they all look back to, but cannot reveal, their origins. Consilience works because Whewell presumes that the human mind is divinely hard-wired to comprehend concepts that are divinely created. Therefore, Whewell can argue that the enquiring human mind is often correct in its intuitive understanding of physical laws, which are then in turn confirmed in scientific terms by empirical evidence.

Whewell's philosophy of science is based essentially on the metaphysical premise that assumes properly conducted scientific investigation eventually leads, through trial and error, to the comprehension of existing general truths. A divine pattern runs through all existence, unifying all, progressing all, and returning all to the original single divine idea. Moving, in other words, from first causes to final causes—this was Whewell's "*one* cycle, yet unfinished". Uniformity appealed to Whewell as a methodology through which consilience might acquire an empirical status. Lyell's "doctrine of absolute uniformity" (*PG*, 1, 87), however, pulls against consilience and teleology by postulating an endless non-progressive system in which the search for origins has no place. Fittingly, Whewell presents his case for the palætiological sciences in the final section of his three-volume *History of the Inductive Sciences* under the heading "History of Geology", the section in which he also discusses the case for and against Lyellian uniformitarianism. His history of geology can be read as a counterbalance to Lyell's "history" and both are exemplary pieces of propaganda. Whewell's categorisation of geology as part of a palætiological group of disciplines that included the moral and intellectual disciplines, as well as language theory, was in distinct contrast to Lyell's strategy of division and his specialised treatment of geology, just as Lyell's strategy of division—his irreconcilable divide between the material world and the human, intellectual sphere—was diametrically opposed to Whewell's vision of consilience and disciplinary cohesiveness.

TENNYSON'S UNIFORMITARIANISM

The debates between Lyell and Whewell lie behind Tennyson's uniformitarian poetics. Tennyson felt Lyell's strategy of division as a rupture in the teleological fabric of contemporary belief systems. Lyell's geology scattered knowledge and fragmented the mind, offering no compensatory sense of unity. Looking back into the distant past Whewell saw the historical disciplines—whether physical or moral—converging; narrowing into a channel that pointed to one single origin. Whewell explains:

> Geology being thus brought into the atmosphere of moral and mental speculations, it may be expected that her investigations of the probable past will share an influence common to them; and that she will not be allowed to point to an origin of her own, a merely physical beginning of things; but that, as she approaches towards such a goal, she will be led to see that it is the origin of many trains of events, the point of convergence of many lines. (*HIS*, III, 484)

It was Tennyson's apprehension of what was at stake in the nuanced differences between Lyell and Whewell—between consilience and disciplinary division—that gave the mid-century poems their radical edge. He grappled with the pull and push of these drives, producing a nuanced critique of the variances between Lyell and Whewell. While he was temperamentally drawn to Whewell's arguments for the palætiological sciences and the expectation of the unity of knowledge, he was nevertheless critical of Whewell's reasoning from a divine premise. At the same time, while Lyell's slick and ostensibly empirical science impressed and sometimes awed him, he saw the ideological fault lines that lie behind Lyell's rigidly imposed uniformitarian laws. *In Memoriam* figures the division that Lyell inflicted on the natural, physical world and the realms of human morality only to find that it shattered the teleological foundations upon which he, like Whewell, based his thinking. The result is an epistemological crisis that once imagined is difficult to repair. However, Tennyson does attempt at least to stabilise the crisis Lyell's uniformitarianism creates and to reinstate teleological purpose in *In Memoriam*. And he does this by superimposing a purposive narrative of linear progress over the non-progressive structure of the poem—a narrative he gleans (as he did for *The Princess*) from Hugh Miller's teleological geology.

Lyellian geology is ubiquitous in *In Memoriam* and *Maud*, from the figuring of eroding landscapes to the uniformitarian shift towards privileging the present over the past—a shift which, in its most extreme case, activates a troubling dialogism. In response to Lyell's strategy of division, *In Memoriam* struggles to reconcile individual consciousness with material nature, while in *Maud*, Lyellian geology triggers the poem's emotional horror and hysteria as Lyell in a calm and commanding style unfolds his paradigmatic uniformitarianism as if its cultural impact were virtually inconsequential. *In Memoriam* and *Maud* both challenge the empirical objectivity that *Principles* claims for itself, showing it to be largely a veneer that once pierced is permanently problematised. The poems see through the attempt to smooth over the rough catastrophist landscape in the postulation of a stable, refined uniformitarian way of thinking. The tensions in the major poems, so often characterised by critics as a conflict between Tennyson's adherence to convention and his propensity towards radical experimentation, are in *In Memoriam* and *Maud* encoded in his inability to reconcile the Whewellian and Lyellian positions. The increasingly centrifugal pull away from the homogeneity of knowledge to disciplinarity marked a shift, not only towards the development of the sciences as we have come to know them, but also in cultural notions about the relationship between the individual and authority and between knowledge and 'truth'. The following chapters hope to show how *In Memoriam* and *Maud* perform those shifts in their uniformitarian poetics as they enact the ideological pull of teleology and the dialogic push of uniformitarianism.

CHAPTER 4

In Memoriam's Uniformitarian Poetics

But I remain'd, whose hopes were dim,
Whose life, whose thoughts were little worth,
To wander on a darken'd earth,
Where all things round me breathed of him
(In Memoriam, LXXXV, 29–32)

The sheer melodrama that was Victorian death is no better envisaged than in G.F. Watts's 1887 painting *Love and Death*. Here, Death,
a shrouded and oddly fleshy, maternal figure, looms in the doorway
of the house of life where Love, a winged and naked boy, falls back in
awe. *Love and Death* and *In Memoriam* (1850) were motivated by the
same experience; just as Tennyson's elegy was inspired by the untimely
death of Arthur Hallam, so Watts painted *Love and Death* after the
early death of his friend the 8th Marquis of Lothian. Indeed, Tennyson
and Watts were themselves friends, although unfortunately there is no
record of Tennyson's thoughts on this highly popular painting, and so
no way of telling whether its sensationalism pushed the bounds of taste
for the ageing poet. In much wider terms, however, the painting does
suggest that by the second half of the century 'death' had survived the
early onslaught of the emerging sciences with all its mysteries still intact.
Geology, which made some of the most startling discoveries of the
early nineteenth century, had as its prime object the study of the vicissitudes of an earth composed largely of dead remains. Rooted originally

© The Author(s) 2017   111
M. Geric, *Tennyson and Geology*, Palgrave Studies in Literature,
Science and Medicine, DOI 10.1007/978-3-319-66110-0_4

in teleological narratives, it seemed almost to promise to unravel the fundamental truths of life and death—to finally reveal that dimly conceived point in time where the marks of divine law could be read unequivocally in material creation. Geology, however, failed to deliver on this front, its deep incursions into the realms of ancient life and death—the fossil record, its postulation of extinction and speculations on deep time and creation—had not succeeded in answering questions about the part that individual death or collective extinction played in the divine plan or even in nature itself. Neither had it succeeded in that greatest of all Romantic quests—the unification of human knowledge as conceived in William Whewell's 'consilience'. Rather, between the publication of Charles Lyell's *Principles of Geology* (1830–1833) and the end of the century, science had begun its multi-divisional breakdown into distinct disciplines and, ostensibly, its long goodbye to culture. By 1887, with Darwin's theories common currency and theology reorganising itself as a separate episteme, Watts was free to mythologise a still inscrutable 'Death' that had resisted the explications of natural philosophy, or indeed 'science', and had failed to be fully assimilated into teleological 'truths'.

A sense of disappointment in the failure of the natural sciences to integrate human meaning stayed with Tennyson all his life. In January 1867, on a visit to Farringford (Tennyson's Isle of Wight home), the Irish poet William Allingham recorded in his diary part of his conversation with Tennyson: "I said I felt happy to-day, but he—'I'm not at all happy—very unhappy.' He spoke of immortality and virtue,—Man's pettiness.—'Sometimes I have a kind of hope'." "Tennyson is unhappy", Allingham notes, because of "his uncertainty regarding the condition and destiny of man" (*Letters ALT*, II, 450–1). The Tennyson that Allingham reveals here seems much less optimistic than the speaker at the end of *In Memoriam* who, after traversing the long and torturous 'way of the soul', is able to affirm that God is love and 'man' ever moves towards perfection in God. Tennyson admitted that his conclusion to *In Memoriam* was "too hopeful...more than I am myself", and seventeen years after the publication of *In Memoriam* it seems he was still drifting with the ebb and flow of faith and doubt—the prospect of immortality set against the unlikeliness of immortality for a presently unworthy humanity.[1] Allingham clarifies Tennyson's disquiet further:

---

[1] Knowles, "Aspects of Tennyson: A Personal Reminiscence" (1893), 182.

His anxiety has always been great to get some real insight into the nature and prospects of the Human Race. He asks every person that seems in the least likely to help him in this, reads every book. When *Vestiges of Creation* appeared he gathered from the talk about it that it came nearer to an explanation than anything before it. T. got the volume and (he said to me), 'I trembled as I cut the leaves. But, alas, neither was satisfaction there'. (*Letters ALT*, II, 450)

Tennyson's trembling anticipation touchingly reveals his remarkable early optimism that *all* might be explained in one tome—that nature and all human understanding might collapse into a nugget of truth expressible in printer's ink and recognisable to the eager truth-seeking mind. Tennyson looked to science for answers to theological and philosophical questions, and he hoped that from his wide reading in these areas and his association with men of science and philosophy he might find a convincing convergence of science and metaphysics that would satisfy his deep need to find spiritual meaning in natural processes, and natural laws in spiritual progress. If Tennyson was disappointed by *Vestiges*, then Lyell's *Principles* was in other respects equally disappointing. *Principles* resonated particularly strongly for Tennyson because its scientific evidence was so impressive by the standards of its day; so much more so than *Vestiges*, which, by contrast, displayed clever conjecture but generalised across a range of disciplines and lacked material evidence for its hypotheses. *Principles* also avoided, on the whole, the clumsy language of theology (an avoidance which would have pleased Tennyson) and closed down the broad speculation and interdisciplinarity that had been a hallmark of so much geological writing.

In Memoriam in some respects can be seen as Tennyson's own attempt to do what Lyell had failed to do—to marry up natural and human law in the revelation of fundamental 'truths'. Tennyson's truth-seeking mind could not ignore Lyell's rigour or the import of his uniformitarian methodologies. In consequence, *In Memoriam* goes much further than simply figuring the implication of *Principles* for a religiously oriented society (readings of this nature observe the metaphoric use of geology and attend to the crisis engendered by *Principles*'s circumvention of scripture). More specifically, *In Memoriam* attempts to provide what *Principles* failed to provide: meaning for itself in human culture. *Principles* was seminal for Tennyson not only because it brought into sharp relief the reality of species extinction in deep time but also because it expounded new ways of thinking about the relationship between the inorganic world and the human world, and an entirely different way of seeing change in the physical world. It was still possible in the 1840s for Tennyson to interrogate death via the patterns of change suggested by geology, and still possible for him to hope that a cultural meaning for death might emerge from the extrapolation of natural laws, if those laws were diligently and precisely traced onto human experience. This is what Tennyson tries to do in *In Memoriam*. Thus, while Watts's vision in *Love and Death* may well have fitted Tennyson's later pessimism and resignation, while he wrote *In Memoriam*, there was still a faint hope that the 'Love' might yet wield a weapon—one forged from natural laws—that could show that there is "wisdom with great Death" (LI, 11) and prove that love indeed is "Creation's final law" (LVI, 14).

Tennyson's investment in poetry suggests a conviction that through its forms the unity of knowledge—scientific, philosophical, social, spiritual—makes itself productively known.[2] As suggested in Chap. 1, in Adamic perceptions there was a tacit sense that the rightness of an idea would become apparent in its expression in language. Thus, in that most exalted and rarefied use of language—the poem—truth would show itself in the cohesion of the aesthetic expression. It is through the poetic plane, in other words, that knowledge, while it is "earthly of the mind", partakes of wisdom, which is "heavenly of the soul" (CXIV, 21–2). Poetic form itself is a vital part of this unity, as while *In Memoriam* self-consciously probes the geological enigmas and loose ends Tennyson found in Lyell's *Principles*, there is a more formal attempt to figure a world circumscribed by Lyellian laws in order to test out whether such a world has poetic validity—to see, in other words, whether Lyell's laws could support human experience and whether teleological purpose could be sustained in the wake of Lyell's science.

---

[2] Different attitudes to the worth of poetry can be seen in the contrast between Thomas Carlyle's and Gladstone's views on Tennyson. Carlyle felt Tennyson's intellect was wasted on poetry, could never understand Tennyson's faith in the power of poetry, as Michael Timko writes: "For Carlyle, Tennyson simply was not a seer, not sincere; he was, instead, a 'Life Guardsman spoiled by poetry'." See *Carlyle and Tennyson* (1988) 57. On the other hand, Hallam Tennyson quotes Gladstone's very generous comment: "Mr. Tennyson's life and labours correspond in point of time as nearly as possible to my own, but Mr. Tennyson's exertions have been on a higher plane of human action than my own. He has worked in a higher field, and his work will be more durable" (*Memoir*, II, 280).

Tennyson's cultural calibration of Lyellian geology was a critical and serious business; as Aidan Day suggests, "Tennyson was not just playing with Lyell and his geology." Day reads Tennyson as finally taking the empirical stance: "*In Memoriam* shows, indeed, that [Tennyson's] imagination, however much it wanted to be able to endorse the spiritually positive vision that he willed into being at the end of the poem, remains addicted to the humanly reductive, critical perceptions of science that had its roots in the rationalism of the Enlightenment."[3] Tennyson was no doubt impressed by the impeccable, gentlemanly forbearance exhibited by Lyell in his measured and confident writing, and possibly (as the poem seems to suggest) overwhelmed by the persuasive force of Lyell's empirical evidence. However, Tennyson was, like his generation, immersed in Whewell's teleological style of thinking, even while he was critical of some Whewell's assumptions.[4] Following Whewell's consilience, he believed "the further science progressed, the more the Unity of Nature, and the purpose hidden behind the cosmic process of matter in motion and changing forms of life, would be apparent" (*Memoir*, I, 323). His continued commitment to the apprehension of "fundamental truths", as suggested by Allingham, may indicate an adherence to "critical perceptions of science" rooted in Enlightenment rationalism, but it also suggests Tennyson's disappointment that science had not yet found what he hoped would be discoverable—an empirical basis for spiritual experience. Tennyson's mind was geared towards a Romantically coloured empiricism, less a Carlylean transcendentalism than a Goethean epistemological outlook based on the belief that spirituality will turn out to have an empirical basis; that science must eventually verify spiritual intuition because (as Whewell thought) the human brain was constructed to discern such ways of knowing.

Following Chap. 3's discussion of Lyell and Whewell, this chapter examines *In Memoriam* as a poem structured by Lyellian uniformitarianism. Isobel Armstrong argues, for example, that Lyell's "modes of 'gradual change in the living creation' are negotiated in the movement of

---

[3] Day, *Tennyson's Scepticism* (2005), 133.

[4] In 1854, for example, Tennyson "carefully studied" Whewell's *Of the Plurality of Worlds* (1853) which argued from a theological basis against the possibility of other planets in the universe being inhabited, concluding; "It is to me anything but a satisfactory book. It is inconceivable that the whole Universe was merely created for us who live in this third-rate planet of a third-rate sun" (*Memoir*, I, 379).

In Memoriam itself, which uses the myth of geology structurally as well as absorbing it into its language".[5] This chapter reads *In Memoriam* as structured on at least four of the basic 'principles' that underpin Lyell's geology. The first of these, and perhaps the most significant, is Lyell's 'strategy of division'—the principle that insists observation of the physical processes by which nature effects change in the organic and inorganic world can have no bearing on matters of human morality or intellect. The second is the principle of 'displacement' and repetition—the idea that the repeated effects of unvarying geological agents operating in deep time suggested continual processes of exchange in the material structure of the earth. The third principle, which supports displacement and repetition, is non-progressive change—Lyell's insistence that change in the natural world offers no evidence whatsoever of progress or purposive direction. The final principle is uniformitarianism itself and Lyell's assumption that only observation of present processes can lead to an accurate understanding of the past—'the present is the key to the past'. These basic principles amounted to ways of understanding geology that contradicted Whewell's perception of teleological purpose and that posed a serious threat to his notion of consilience. For *In Memoriam*, however, these principles form the logic and ordering structures that shape the poem's movement. They provide fitting patterns for the expression of unending grief, for the discursive examination of faith and doubt and for exploring the relationship of individual death to the wider scheme of nature. Deploying Lyell's material laws poetically provides a landscape on which to stage the poem's ruminations on Hallam's physical remains; geological laws describe the fate of these precious remains and the geological structure of the poem becomes a material/formal landscape where the speaker is able to remain with Hallam's remains.

The staging of human emotion on a Lyellian landscape, however, has negative consequences, as for a mind accustomed to perceiving teleological purpose, Lyell's principles offer little spiritual sustenance. Rather they establish a profoundly troubling, decentred material world. Delivered onto this uniformitarian poetic landscape, the speaker is subject to the logic of its principles and its laws of movement, which are immutable and inescapable. While the poem's structure works persuasively to express unending grief, it inevitably compromises the elegiac expectation of recovery. As the

---

[5] Armstrong, *Victorian Poetry: Poetry, Poetics and Politics* (1993), 252.

laws of change are invariable, the earth is envisaged as moving through time but having no discernible beginning or end—thus, the Lyellian earth has no teleological direction. Such a pattern of movement produces in the poem a speaker who has lost his connection with the past and his narrative of self. This loss of narrative direction is intrinsic to the poem's fragmented form and, similarly to Lyell's seemingly infinitely processing system, it has consequences for closure. The poem itself, of course, must end, and to bring it to anything like a satisfactory conclusion a thematic narrative of progress sympathetic to Whewell's sense of consilience and drawn in part from Hugh Miller's geology is superimposed, however unconvincingly, over *In Memoriam*'s massive non-progressive Lyellian structure.

LYELL'S STRATEGY OF DIVISION

Lyell's 'strategy of division' in *Principles* was partly a rhetorical manipulation designed to abolish the reader's anthropocentric perception of time in order to bring uniformitarianism into relief. Roy Porter suggests that "For Lyell, so many theories of the earth had been scientifically useless because the earth had been conceived anthropocentrically and anthropomorphically. Man's hopes and fears, his pride and guilt, his thoughts about origins and destiny, had been projected onto the earth."[6] Lyell's aim was to demonstrate that an accurate understanding of geology could only be achieved when the earth was viewed, as far as was possible, from outside human time and human perception—a position that, by default, reversed assumptions that earth had been created solely for the purpose of accommodating 'man'. As Chap. 3 argued, Lyell's division was also instituted in order to offset any imputation of evolutionary theory that deep geological time might encourage. It was put in place in order to account for the appearance of humanity as a special event, while at the same time avoiding any inference of progress or unprecedented occurrences in nature that would compromise his postulation of unvarying uniformitarian laws. Thus, Lyell asserted that "the earth's becoming at a particular period the residence of human beings, was an era in the moral, not in the physical world" (*PG*, I, 163) and therefore was an occurrence outside the regular, uniform laws that operate across physical nature.

---

[6] Porter, "Charles Lyell and the Principles of the History of Geology", *British Journal for the History of Science*, 9 (1976), 93.

Michael Tomko argues that Tennyson took up Lyell's division in *In Memoriam* because both men were "engaged in similar, not contradictory, projects that revise Paleyan natural theology into a dynamic spiritualism that draws on an absolute and prophylactic barrier between the physical and spiritual world" (113–4). Lyell's insistence on division enables him to "reorient…belief into a spiritualism in accord with scientific ideology" (114). Tennyson, Tomko believes, abides by this division (albeit reluctantly at first because it meant relinquishing Hallam's physical remains) in order to perform in the poem a similar conversion into spiritualism that would allow the speaker to finally take solace instead in the prospect of a spiritual reunion with Hallam.[7] This, however, does not take into account that Lyell's division was, as argued in Chap. 3, a strategy and therefore not specifically formulated for spiritual reasons. Lyell's division is a key sticking point for Tennyson because it denied the possibility of a Whewellian perception of the unity of knowledge in which the 'fundamental truths' of nature and the human world could be seen as harmonious—a unity that, according to Allingham, Tennyson was still hoping to find in 1887. Lyell's strategy of division may have at first seemed to offer a way of unfixing faith from a careless and indifferent nature—the fickle "Nature" of *In Memoriam*'s famous sections LV and LVI, who appeared to be not only "careless of the single life" (5–8) but, on the geological evidence of "scarped cliff and quarried stone", careless of all life; "'A thousand types are gone: / I care for nothing, all shall go" (2–4). However, the implications of seeing nature as indifferent to humanity results in the "evil dream" that "God and Nature" are "at strife" (LV, 5, 6).

The poem's enactment of Lyell's division is part experiment—what would it be like to experience a world based on Lyell's division?—and part an act of debunking that division. Tennyson understood the rhetoric behind Lyell's insistence on division and he felt its clumsiness in what was largely an empirical and persuasive argument. If the poem's aim was to forge a spiritual pathway that could run parallel but never intersect

---

[7]Tomko draws on J.M.I. Klaver's *Geology and Religious Sentiment* for his reading of Lyell. Klaver suggests Lyell had a "predilection for Unitarianism" which has been largely overlooked and that Lyell would have been attracted to its "progressive outlook". See Klaver (1997), xiv, 78.

with the trajectory of material science, then the poem succeeds on this level, and does so because it operates according to the divisions that Lyell himself makes. However, Tennyson's continued commitment to the unity of knowledge and his Romantic attachment to nature meant that he could not profit from a spirituality divorced from nature's forms and patterns. Lyell's division may have allowed some gains in loosening ties with an inflexible religious doctrine and Scriptural literalism, but the cost of this nonconformity—a uniformitarian landscape emptied of spiritual meaning and purpose—was for Tennyson ultimately too great.

DISPLACEMENT AND REPETITION

Despite the difficulties Lyell's strategy of division posed, other patterns of change expounded in *Principles* were sympathetic to *In Memoriam*'s thematic expression. For example, Tennyson discovered in *Principles* the virtue of repetition. Lyell's uniformitarianism was a premise from which to argue 'scientifically' that the laws of geological change should be assumed to be unvarying.[8] Lyell encouraged a perception of change as uniform across time, and a sense of the repetition of types of actions pervades *Principles*. He used repetition to his advantage; his aim was to convert readers to a uniformitarian cast of mind via the replication and reinforcement of material evidence, and to alert them to the profound effects that can be wrought from geological processes operating in the present. James Secord argues that "The sheer length of the *Principles*—over 1400 pages in the original edition—was essential to this programme

---

[8]Hallam Tennyson acknowledges that his father was "deeply immersed in ... Lyell's *Principles*" in 1837 (*Memoir*, I, 162). It is generally agreed that the first sections to be written, and those that definitely pre-date Tennyson's reading of *Principles*, were XXX, IX, XVII, XVIII, XXXI-XXXII, LXXXV and XXVIII. See, for example, Christopher Ricks, *Tennyson* (1972), 121, Susan Shatto and Marion Shaw, eds., *Tennyson: In Memoriam* (1982), 6. Shatto and Shaw suggest that dating the composition of the lyrics is particularly difficult because "Tennyson was in the habit of re-using old notebooks, and two adjacent poems may in fact have been composed at widely different times" (7). However, there is evidence that he was reading Lyell earlier in 1836, as a letter to Richard Monckton Milnes attests, where he refers to Lyell's theory of climate change as given in chapter eighteen, volume two of *Principles* (*Letters ALT*, I, 145). More significantly, Dennis Dean demonstrates that Tennyson was familiar with Lyellian ideas much earlier than 1836 from his reading of his "favorite periodical", the *Quarterly Review* (*Tennyson and Geology*, 1985, 5).

of perceptual reform." Lyell used a rich array of observed data from many diverse sources, as well as literary references, all designed to overcome resistance to uniformitarian theorising. Thus, the "calmative effect of hundreds of examples, made readers into witnesses to the power of modern changes".[9] Readers were gradually converted to a uniformitarian perspective as the evidence mounted and uniformitarian thinking itself became an established way of seeing and knowing in the minds of readers. Just as geological processes operated imperceptibly to change the entire landscape, diligent readers would be worn down—subtly eroded by wave after wave of evidence while simultaneously a new perspective gradually arose in them—thus, they would find themselves, by the end of the three volumes, inhabiting an entirely new conceptual landscape. And this is certainly the way Darwin famously read Lyell as "alter[ing] the whole tone of one's mind".[10] Fittingly, the major geological processes that Lyell advocated in *Principles* (in order to demonstrate that present processes operating in deep time were sufficient to account for all past change) were, like his rhetoric, largely slow acting and accumulative. Thus, there was an elegant and powerful symmetry in Lyell's science and rhetorical strategy—the former offering an understanding of geological change through the accumulative effects of processes presently visible, the latter attempting to gradually erode readers' received perceptions through the accumulation of evidence for uniformitarian change.

In similar fashion, repetition becomes the major pattern for the expression of the speaker's out-of-proportion grief.[11] *In Memoriam*'s abba-rhymed quatrains work to extend indefinitely the grief that a more conventional elegy, in its function as a paradigm for recovery, specifically seeks to end. Where the use of quatrains becomes distinctive, even eccentric, is in terms of quantity; over seven hundred such quatrains, all unvarying in their metre and rhyme scheme, is unprecedented and almost alarming in its persistence. Such rigid and uniform repetition gestures towards the interminable—towards Lyell's fixed laws of

---

[9] Secord, *Principles* (1997), xx.

[10] Darwin to Leonard Horner, August 29, 1844, *Correspondence of Charles Darwin*, 3 vols., Fredrick Burkhardt and Sydney Smith, eds. (1987) III, 55.

[11] A. Dwight Culler, observing connections between Lyellian geology and *In Memoriam*, notes that "if we think that Tennyson requires a long time to move from grief to reconciliation, we should consider what Lyell required for the formation of mountains and seas." *The Poetry of Tennyson* (1977), 150.

change operating against a backdrop of geological time. So small a unit as the quatrain seems to be absurdly extended beyond what it might be expected to sustain, and in this it shares with Lyell's uniform laws a sense of the power of small actions in extended time. On a wider level, the poem produces its effect via gathered evidence, mobilising one after another numerous scenarios and vignettes that assess and reassess the power of love and the meaning of death, the effects of doubt and the quality of faith, that work to form and reform the speaker's experience of grief, continually keeping it present. Calmative evidence allows readers to view the speaker's emotion from various angles, and makes them witness to a grief so powerfully present that it has the potential to reshape perceptions of the past. Thus, as Lyell assembled evidence to argue that the present allows us to understand the past, *In Memoriam*'s accumulated lyrics make present grief the arbiter of a past, now seen entirely differently in the light of loss, as in section XXIV discussed later, where a speaker specifically located in the present wonders if the past really was "As pure and perfect as I say?" (2). Thus, the speaker is converted to uniformitarian thinking not only in terms of the structure of Lyell's argument—the repetition and accumulation of evidence—but also in Lyell's methodological approach which sees the 'present as the key to the past'.

The poem's length has always been an issue. Tennyson famously asserted that the *In Memoriam* lyrics were not written "with a view to weaving them into a whole, or for publication"; it was only after he found that he "had written so many" lyrics around his grief for Hallam that he felt they might be brought together as a single poem (*Memoir*, I, 304), which suggests that there was no premeditated linear structuring of the poem. *In Memoriam* seems to have developed from the initial resolve to write about the loss of Hallam in a particular form; as Michael Mason suggests, "Tennyson's decision, whenever it occurred, to reserve the special ABBA stanza for all poems on Hallam's death, and only those poems, must have followed on some kind of plan to assemble these as a group."[12] Christopher Ricks argues that the poem's proposed title, *Fragments of an Elegy* implies "with frankness and probably with truth, that the poem as a whole does not possess a firm focus". The fitness of *In Memoriam*'s form, however, lies in part in its fragmentation, as the

---

[12] Mason, "The Timing of *In Memoriam*" in *Studies in Tennyson*, Hallam Tennyson ed. (1981), 60.

lyrics capture the small units of feeling that surface in the mourner; the short stanzas and brief lyrics match the grieving experience, the mind interrupted from the demands of routine life by memory, the intervals of forgetfulness pierced by the sharp remembrance of loss. For Ricks, "the most important critical question about *In Memoriam* remains the first and most obvious one: in what sense do the 133 separate sections, ranging in length from 12 lines to 144 lines, constitute a whole, a poetic unity, a poem?"[13] The compulsion to keep producing elegiac lyrics, however, seems to have outstripped concerns for the kind of unity Ricks expects.[14] However, the poem's stanzaic form itself seems to offer a unity of kinds, as Sarah Gates suggests, the stanza form is the "only constant—and an obsessive constant it is—to be found" in the poem.[15] Tennyson's knowledge of Lyellian geology also implies a rationale for seeing unity in ever-repeating stanzas. Already versed in the uniformitarian perspective, on reading Lyell's full text, Tennyson was exposed to the formal cohesion of *Principles* indicated by uniformitarian laws that figure the natural, consistent and repetitive patterns against which human concerns play out. The formal structure of *In Memoriam* constitutes the level at which these laws rule human perception and circumscribe the material limits of human experience. The completed poem emerged from the gathering up and ordering of the lyric sections originally unfixed from any particular sequence.[16] The final version, its length, its repetitive quatrains

---

[13] Ricks, *Tennyson* (1972), 212.

[14] T.S. Eliot found *In Memoriam* to have "only the unity and continuity of a diary" but a diary, nevertheless, "of which we have to read every word", while Eric Griffiths suggests "*In Memoriam*'s unity principally stems from a feature which would be surprising if ever found in a diary: it is written throughout in the same stanza." Eliot, *Essays Ancient and Modern* (1932), 196. Griffiths, "Tennyson's Idle Tears" in *Tennyson Seven Essays*, Philip Collins ed. (1992), 46.

[15] Sarah Gates, "Poetics, Metaphysics, Genre: The Stanza Form of *In Memoriam*," (1999), 508.

[16] Sections, XXX, IX, XVII, XVIII, XXXI-XXXII, LXXXV and XXVIII, Christopher Ricks points out, were of a type: "The striking thing about this early group is that it evinces the less perturbed calm which *In Memoriam* mostly intimates to be an achievement slowly won rather than immediately entered upon." *Tennyson* (1972), 121.

and unwavering abba rhyme scheme, feel entirely fitting, as many critics have suggested, for the expression of a mental state in which the psyche is stuck within the trammels of grief.

The discursive microdialogue of the poem, however, and its movement between faith and doubt also encodes Lyellian patterns of displacement. Like James Hutton before him, Lyell's aim was to suggest an overall harmony in the system of nature. On the surface, this aim would have been attractive to Tennyson. The earth, according to Lyell, is governed by strict uniformitarian laws that preserve its delicate harmony. Lyell spends the largest part of volume one of his *Principles* offering evidence for the far-reaching effects of igneous and aqueous geological forces working in unison. His aim is to show that there is a balance in these operations that keeps the ratio of land and sea constant. For example, in his treatment of aqueous agents, Lyell writes:

> The sediment carried into the depths of the sea by rivers, tides, and currents, tends to diminish the height of the land; but, on the other hand, it tends, in a degree, to augment the height of the ocean, since water, equal in volume to the matter carried in, is displaced. The mean distance, therefore, of the surface, whether occupied by land or water from the centre of the earth, remains unchanged by the action of rivers, tides, and currents. (*PG*, I, 475)

Erosion and sedimentation describe a process of 'displacement' in which the same materials appear continually to exchange places. Change occurs, but it is non-progressive change that keeps the system in balance. Surface land is eroded or destroyed in one place, only for there to be a corresponding uplift and re-formation elsewhere in the system. Thus, the earth shifts continually in the repeated play between ruin and renewal and, paradoxically, everything is in continual flux, constantly changing places, while the wider picture suggests that everything "remains unchanged". Similarly, faith and doubt, hope and despair shift continually in the poem as the speaker moves from the heights of meaning, for example, in the ecstatic lyric of CXV when Hallam's "living soul was flash'd on mine" (36), into the troughs of despondency when faith is "cancell'd, stricken thro' with doubt" (44). Displacement is the rule in all the movements between faith and doubt in the poem; what can never be achieved while this formal pattern prevails is a dialectic synthesis that might activate progress. An acknowledgement of the significance of displacement is found early in the poem, where the speaker wonders if there can be a corresponding gain in loss; if we can "find in loss a gain to match?" (I, 6).

In the shifting landscape of Lyell's system of displacement, however, there is only ever a finite quantity of matter in the system. The earth shapes and reshapes itself continually in the present, and there is something very poignant about the process of displacement in which the same materials are continually rearranged, whether geological materials, or the finite gamut of signifiers that constitutes *In Memoriam*'s language of love and loss, as they foreground respectively the limits of physical existence and the insufficiency of language. The speaker searches the bounds of these finite systems in the agonising quest for a certainty that neither can offer, as nothing can be added or taken away, no progress or gains can be made in the closed system of continual displacement. The continual forming and re-forming of finite materials in displacement helped to envisage Hallam's physical self—his remains—as permanently remaining. However, they are only available in a form now teasingly unsatisfactory. The loved other is "turn'd to something strange" (XLI, 5) and what remains must enter the geological system of displacement which is the fate of all remains. From here there can be no reconstitution of remains, no return to a former state, however much lyric themes rehearse those possibilities. Sifting through the vestiges of remains—remembered touches, gestures, words, letters—in Lyell's humanly unmeaning landscape, is merely a continual attempt to reconstruct the same materials in different but always equally inadequate forms.

Repetition is a well-documented literary device in *In Memoriam*. There is regular use of anaphora, for example: "Thou madest" and "Our wills" in the opening section; "Peace and goodwill, goodwill and peace, / Peace and goodwill" (XXVII), and "Ring out" and "Ring in" of CIV to name only a few. And the doubling of words: "hand-in-hand", "each at each", "orb to orb", "veil to veil", "Rise, happy morn, rise, holy morn" in section XXX alone, with other examples too numerous to mention.[17] Repetition evokes a sense of being outside linear time. For example, in section XXII and XXIII the past spent with Hallam is envisaged as a complete and unchanging past:

---

[17] On repetition see, particularly, Alan Sinfield, 1986, 114.

> The path by which we twain did go,
> Which led by tracts that pleased us well,
> Thro' four sweet years arose and fell,
> From flower to flower, from snow to snow:
> And we with singing cheer'd the way,
> And, crown'd with all the season lent,
> From April on to April went,
> And glad at heart from May to May. (XXII, 1–8)

While the seasons may change, the doubling of words foregrounds sameness as they move "From flower to flower, from snow to snow", "April on to April", "May to May" (4, 7–8). Paradoxically, change appears to effect no change whatsoever. However, the next lines describe how a catastrophic but unforeseen change—Hallam's forthcoming death—lurks in the future:

> But where the path we walk'd began
> To slant the fifth autumnal slope,
> As we descended following Hope,
> There sat the Shadow fear'd of man;

The 'descent' towards death is marked by a shift in the landscape as the "path we walk'd" "slant[s]" downwards. The unchanging seasons, thus, belie a deeper imperceptibly change occurring in the landscape as larger movements of displacement range below seasonal adjustment, effecting much broader and far-reaching change. The path, which represents the passage through time, 'descends', taking everything that moves through time further into the dark geological world of dead matter where meaning is lost in an eternity of repetitive processes of displacement. By the next section the speaker, "looking back to whence I came / Or on to where the pathway leads", finds that all has changed entirely: "How changed from where it ran / Thro' lands where not a leaf was dumb" (XXIII, 7–10). Figured by uniformitarian patterns of displacement in which the material landscape shifts almost imperceptibly on a global scale, Hallam's death is nevertheless envisaged as a transformation of geological proportions, one that delivers the speaker onto an alien landscape that has been struck dumb, having been emptied out of human meaning by Lyell's strategy of division. The change is suitably subtle as while it represents catastrophe in the human world, in Lyell's material world it is no change at all.

Non-Progressive Change

no vestige of a beginning,—no prospect of an end[18]

While patterns of displacement suggest repetition, Lyell's geology was also notoriously non-progressive. Like his uniformitarian predecessor, James Hutton, Lyell did not see geology as concerning itself with beginnings and ends. The uniform laws of nature did not vary and it was useless to look for the marks of creation in the geological processes of the earth. Lyell's geology was, as James Secord terms it, "profoundly ahistorical".[19] And as with Lyell's strategy of division, it was imperative to argue for a non-progressive system not only in order to avoid the language of progress that might encourage theories of evolution, but also to avoid any imputation that the laws of geological change might have operated at different rates or been of different types in the past. Deep time, coupled with unvarying laws of change, implied a hardly imaginable sense of purposeless repetition. Lyell writes with reference to Hutton's uniformitarianism: "The imagination was first fatigued and overpowered by endeavouring to conceive the immensity of time required for the annihilation of whole continents by so insensible a process. Yet when the thoughts had wandered through these interminable periods, no resting place was assigned in the remotest distance" (PG, I, 63). No matter how far the geologist looked back into the strata, no signs of first causes could be found. In Lyell's geology, there was only non-progressive change in infinite process. This describes *In Memoriam*'s form, with its enclosed quatrains that act out the repetitive actions of displacement imparting a sense of the rise and fall and the ebb and flow of states of feeling. Change occurs with no clear sense of progression, just as in Lyell's material system, continuous change plays out with no prospect of progress.[20]

Non-progressive change is also suggested via the poem's repetitive stanzaic form, which allows movement while continually returning to the same

---

[18] Hutton, *The Theory of the Earth* (1795) I, 200.

[19] Secord, *Principles* (1997), xviii.

[20] For the critical accounts of *In Memoriam*'s stanzaic movement, see Alan Sinfield (repetition with difference) and Sarah Gates (spiral movements) in terms of theme and form respectively (Sinfield 1986; Gates 1999).

condition of emotional being. For example, where the speaker expresses his ambivalence concerning recovery; his desire to move on against his desire to remain in the landscape where Hallam's remains remain:

> Still onward winds the dreary way;
> I with it; for I long to prove
> No lapse of moons can canker Love,
> Whatever fickle tongues may say. (XXVI, 1–4)

The abba rhyme scheme is the uniform constant that represents a love that cannot change. The poem's form 'proves' the point as there is "no lapse" in the "dreary way" that must "wind" seemingly indeterminably "onward" in order to supply 'proof' of the unchanging nature of love.[21] The abba rhyme scheme helps to set a Lyellian pattern suggestive of displacement, repetition and non-progression instilling a fixity of form that guarantees a continual return to the same pitch of feeling. The later phrase, "With weary steps I loiter on" (XXXVIII, 1), like the paradox of "still onward", again encodes the desire to prove through change, that nothing changes. Tennyson is a master at manifesting the vacillating mind, as Gregory Tate attests; "while the minds of *In Memoriam* and *Maud* do not straightforwardly progress, they change frequently, even though they appear to be at times pathologically fixed in one emotional state or idea". Tate notes the strange "sense of a simultaneous permanence and successiveness of the psyche [that] pervades *In Memoriam*".[22] The form that this paradox takes is Lyellian; that is, as in Lyell's geology, consistency is proved by change.

THE PRESENT IS THE KEY TO THE PAST

A much-neglected aspect of Lyell's uniformitarianism in terms of its meaning for culture is the postulation that present causes explain the past. Lyell effectively asked readers to privilege the present over the past—to see the past as the consequence of processes observable in the present. This methodology gave Lyell's geology an empirical edge that

---

[21] Gates reads this pattern as a spiral movement; the "second 'a' 'returns,' but it also leads 'beyond' because it is different from the middle couplets and only faintly recollects its partner" (1999, 508).

[22] Tate, *The Poet's Mind* (2012), 96.

seemed to usher in a 'properly scientific' way of thinking. However, Lyell used this approach not merely as a methodology but, as Chap. 3 explained, as a substantive law of geological nature. To sensibilities inclined to imagine natural laws as descriptive of moral and metaphysical 'truths'—Whewellian and Trenchian teleological sensibilities—the shift in emphasis from the past to the present had radical implications. Uniformitarianism reversed Adamic and teleological thinking; it flew in the face of Adamic theory and the conservative emphasis on the authority and superiority of the past. Where *The Princess* established conservative ideologies of class stratification and gender via the examples of the past—Miller's "master existences" that offered ostensibly non-linguistic proof of an order immanent in all nature—Lyell's geology shifted the focus away from the past and onto the present.

Thomas Carlyle's *Past and Present* (1843) offers a rarefied and idiosyncratic example of received perceptions of the hierarchy of the past over the present. *Past and Present* embodies that hierarchy in its title and method. The book is designed to repair a morally degenerate present via the example of the past in order to steer an ideologically sound path into the future. *The Princess* used a similar pattern; the aberrations of the present are resolved by reference to a past (medieval, geologic) which allows the future to proceed according to the authority of the past. Carlyle argued with typical force that a "godless" 'Present' could not access a 'truthful' 'Past' and could not therefore progress itself:

> To predict the Future, to manage the Present, would not be so impossible, had not the Past been so sacrilegiously mishandled; effaced, and what is worse, defaced! The Past cannot be seen; the Past, looked at through the medium of "Philosophical History" in these times, cannot even be not seen: it is misseen; affirmed to have existed,—and to have been a godless Impossibility. ... For in truth, the eye sees in all things "what it brought with it the means of seeing." A godless century, looking back on centuries that were godly, produces portraitures more miraculous than any other. All was inane discord in the Past; brute force bore rule everywhere; Stupidity, savage Unreason, fitter for Bedlam than for a human World![23]

---

[23]Carlyle, *Past and Present*, 297–299.

As a 'key to the past', in Carlyle's thinking, the present (and for him a very specific present) inevitably distorts what it sees; it leads 'men', in Lyellian fashion, to see the past as operating via the same processes that operate in the present: "How shall the poor 'Philosophical Historian' to whom his own century is all godless, see any God in other centuries?" It must be "taught again" that "It is…not true that men ever lived by Delirium, Hypocrisy, Injustice, or any form of Unreason, since they came to inhabit this Planet". By recognising the primacy of the 'Past', regardless of how different it might appear from the present, men's "acted History will then again be a Heroism; their written History, what it was once, an Epic". For Carlyle and for an age of medieval pretensions and Gothic revivalism, the past most emphatically was a vital key to the present. It alone afforded a sense of continuity, order and identity in an age of constant change. For teleological thinkers such as Whewell, who recognised how uniformitarianism's emphasis on the present had a powerful potential to strengthen the claims of all narratives, biological, social and political, to validate change in the present, to see the present as the key to the past was a subtle radicalism of the first order. To substitute the authority of the past with that of the present was to assault the fixed foundation of the social order, national identity and the authority of God.

Added to this, the prioritising of the present deflected from teleology's concern with purposive direction, foregrounding instead a sense of open-endedness. Fixed in the present and ignoring first and final causes, Lyell postulated a physical earth without a narrative sense of purpose or direction, and therefore, one that can never, at least on the uniformitarian evidence, be itself a complete and finished whole. Similarly, the 'I' of *In Memoriam* is a uniformitarian 'I', itself always unfinished and permanently deprived of the hope for wholeness. It is, as many critics have asserted, a mind in process. Gregory Tate argues, for example, that *In Memoriam* "is not just a lyric expression of thought and feeling but also a staged, self-conscious, and self-questioning critique of its speaker's mental processes".[24] This "psychological stance also operates on the level of the poem's formal structure. The mental processes that comprise *In*

---

[24] Ibid., 96.

Memoriam are presented in dozens of discrete poetic fragments and pal-inodes which enact its account of a mutable mind."[25] The poem loses its linearity in discursions and repetitions that mimic not only the repetitive actions of geological displacement but the processes of consciousness. A. Dwight Culler makes a connection between the poem's elegiac form and geology, suggesting that "Tennyson in his mature poetry is gradual-ist rather than catastrophic in his assumptions. The pastoral elegy, on the other hand, is an apocalyptic or catastrophic form. Depending as it does on sudden revelation or 'discovery' that the beloved is not dead but in some sense lives on." The elegy's revelation would not normally unfold in real time, rather, "one has the feeling the 'discovery' was planned from the very beginning and that the whole exercise takes place artificially in space", whereas in *In Memoriam* it takes "place in time".[26] However, tak-ing this further, the consciousness that pervades the poem could be said to be not in an 'artificial space' or 'in time' but present in the moment, present in the processes of consciousness that appear to occur in the present. This is a mind that does not move through time but that—in Lyellian fashion—is changeable in the present. The psychological condi-tions created in the poem have all the marks of Lyellian ways of perceiv-ing. Just as in prioritising the present over the past the earth's teleological narrative is lost, the elegy's narrative flow is dissolved by its indetermi-nable Lyellian form. The effect is to destabilise the path (the begin-ning, middle and end) of elegiac recovery. *In Memoriam*'s length, its repetitions, its model of displacement and its depiction of the speaker's consciousness as a present consciousness are all, at base, Lyellian and uni-formitarian and the uniformitarian quality of the poem produces one of Tennyson's most radical poetic effects.

---

[25] Tate (2012), 96. F.E.L. Priestley argued that the extended length of the elegy plays an important part in the creation of consciousness, as "To extend the time-scale is to bring the movements of emotion closer to the actual, and to explore them in greater detail [...] The linear pattern of the formal elegy is thus replaced by a highly complex one, in which the clear succession of movements is blurred by overlappings, interfusions, anticipations, reflec-tions and retrospects, by movements set against movements, and so on" (F.E.L. Priestley, *Language and Structure in Tennyson's Poetry*, 1973, 122).

[26] Culler, *The Poetry of Tennyson* (1977), 150.

The Uniformitarian 'I'

The only anchor the speaker's present consciousness appears to have to keep his narrative in touch with time is found in the three Christmases. Without this hook into time, the repetitive abba stanzas would lose any intimation of linearity and the poem would fail as an elegy, as there would be no possibility of movement into recovery from grief. The three Christmases pin the free-floating lyrics to a timeline, yet they feel tacked onto what seems otherwise to be an arbitrary arrangement of stanzas and sections. They foreground how human time is projected onto the seemingly purposeless movements of the earth. Without this timeline, the speaker would be set adrift on the poem's waves of repetitive movement. However, despite the timeline of the three Christmases, in times of doubt the speaker experiences his loss of narrative flow acutely, as in the "Dark house" lyric, when the "noise of life" seems "far away" and the unmarked, nameless "blank day" "breaks" on the "bald street" (VII, 9–2). At these moments, the speaker rubs up against the existential remoteness of Lyell's material world and its laws of change. These laws have churned out unnumbered 'blank days' since time immemorial. Days that have rolled past unregistered by any human consciousness, and that will continue to do so long after the humanly constructed world of meaning—"where the long street roars"—is once more recycled in mammoth movements of displacement into the humanly inhospitable "stillness of the central sea" (CXXIII, 3–4).

The three Christmases help to fix the speaker's sense of self, as without them he would be stuck in the uniformitarian present ostensibly forever sifting through the finite materials of a geologically and linguistically non-progressive landscape. The timeline aids the fabrication of selfhood by marking the past via memory and in the expectation of the future. Working against the non-progressive form of the poem, the Christmases chop up time into past and future, diverting consciousness from its condition of being in the present. They help to formulate a sense of self as in transit through time, accumulating selves and anticipating future selves in the realisation of what is assumed to add up to an ultimate whole. They placate what is otherwise the speaker's painful recognition that such a whole self might not be available at all; just as a teleological knowledge of creation's first and final causes is unknowable in Lyell's geology. They help the speaker to escape the workings of his own mind in the present and to escape his debilitating sense of the inexorable nowness of consciousness and its import that this nowness might be the sum total of the self.

The speaker's sense of self owes much to Arthur Hallam's philosophical writing.[27] For Hallam, the 'soul', which he genders female (as Tennyson also does in his poem "The Palace of Art") is permanent, although experienced by consciousness as in parts. In "On Sympathy", Hallam discusses the soul, suggesting "It is the ultimate fact of consciousness, that the soul exists as one subject in various successive states...Far back as memory can carry us, or far forward as anticipation can travel unrestrained, the remembered state in the one case, and the imagined one in the other, are forms of self."[28] The soul—the self's wholeness and identity—is "one subject" in "successive states"; this is a condition of being. Memory, as well as our imagined future selves, make up successive states that guarantee the completion of the whole soul/self even while it is unfolding. Thus, Hallam posits the past and also the future as creative of the whole, suggesting an intimate teleology of self that mirrors larger collective teleological expectations. Hugh Miller's geological thinking, as addressed in Chap. 2, works with the same teleological patterns. Where Hallam saw the soul as composed of successive selves, including past and future, Miller saw creation as made up of all the successive formations of the earth. The geologist must unite "the links of the chain of creation into an unbroken whole" as "The perfection of the works of Deity is a perfection entire in its components, and yet these are not contemporaneous, but successive: it is a perfection which includes the dead as well as the living, and bears relation in its completeness, not to time, but to eternity." This is the macro version of Hallam's "one subject" in "successive states" in which, in Miller's geological terms, "we find the present incomplete without the past" (*ORS*, 45). *In Memoriam*'s uniformitarian 'I', however, subject to the uniformitarian laws inscribed in the poem's form, is a self trapped in a present in which in Hutton's uniformitarian terms there is "no vestige of a beginning, no prospect of an end", where wholeness and ultimate transcendence is unattainable. Paradoxically, it is the speaker's reluctance to relinquish a material world in which "my lost

---

[27] Eric Griffiths demonstrates the importance of Hallam's Kantian-orientated philosophy for *The Princess*'s lyric "Tears Idle Tears". Gregory Tate also usefully draws attention to Hallam's influence on the creation of the self in *In Memoriam*. See, Eric Griffiths, "Tennyson's Tears Idle Tears," in *Tennyson Seven Essays*, Philip Collins ed. (1992). Gregory Tate, *The Poet's Mind* (2012), 96.

[28] Arthur Hallam "On Sympathy" in *Remains in Verse and Prose of Arthur Henry Hallam* (1834), 100. Hereafter cited as *Remains* parenthetically.

Arthur's loved remains" (XI, 3) still remain, that lies behind the unattainability of wholeness.[29] This is the problem of section XLVII, where the speaker realises that to attain wholeness is to give up desire, to lose his individuality in the "general soul" (4). The success of *In Memoriam*'s non-progressive form to express unending grief and desire is thus also what stalls the self's development and keeps it in a non-progressive, and continually processive, present consciousness.

According to Hallam's own metaphysics of the soul, it is through recognition of the 'other' that the self comes into being. In "On Sympathy", the soul (which, like the Keatsian 'poetical character' takes into itself the 'other') achieves wholeness by gathering up its successive selves, which come into being through the self's recognition of the other.[30] Hallam explains: to 'become', "to add a mode of being to those we have experienced", is the natural pleasure of the soul, thus, "through the desire to feel as others feel, we may come to feel so" (*Remains*, 99). In this way, Hallam theorised a type of Hegelian realisation of spirit in the separation of the self from its surroundings and the subsequent assimilation of the other into the self as stages of the development of the soul. The self is separate and whole simultaneously, finding itself and its self-realisation in the other. The discovery of another consciousness as "having a world of feeling like the soul's own world" produces the question: "how can the soul imagine feeling which is not its own?" (*Remains*, 100). The soul solves this "conception, only by considering the other being a separate part of the self, a state of her own consciousness existing apart from the present, just as imagined states exist in the future. Thus absorbing… this other being into her universal nature, the soul transfers at once her own feelings, and adopts those of the newcomer" (*Remains*, 101). It is a dialectic that foregrounds how the speaker's stasis pivots on the unavailability of the other to effect the dialectic process. The reciprocal sense of self, however, also acknowledges the dialogic self—the self constructed in the recognition of the other. Michael Holquist describes the dialogic self as both self-created—"we *must*, we *all* must, create ourselves, for the self is not given"—and as limited by "the material available for creation"

---

[29] Devin Griffiths writes innovatively on the influence of Hallam's writings—particularly his poetry—on *In Memoriam*, arguing that the poem tries to "weave Hallam back into the family", *The Age of Analogy* (2016), 130.

[30] Peter McDonald offers a nuanced reading of "On Sympathy" as an essay on Wordsworth. See *Sound Intentions* (2012), 197.

because these materials are "always provided by the other".[31] The problem for the speaker is that without Hallam not only is he denied the synthesis required for development towards wholeness, more fundamentally, he has lost the sense of himself as dialogically performed—as coming into being in the interaction of self and other; the moment imagined as the "endless feast" (XLVII, 9). This is a distinctly dialogic moment of exchange but not synthesis. Hallam's missing presence deprives the speaker not only of his reciprocal sense of self-development—the speaker is "No more partaker of thy change" (XLI, 8)—but of his experience of self, as the other's vision is the vital element in our perception of ourselves. The speaker has lost the linear narrative of self sustained by the other; he has lost the link to past selves and the prospect of future selves, and exists only in non-progressive present consciousness.

The self of *In Memoriam* is a uniformitarian self in the same sense that Lyell's uniformitarian story of the earth was "profoundly ahistorical".[32] For Sarah Gates, the speaker of *In Memoriam* is peculiar because he is not a speaker who looks back in order to narrate a history of the past, but rather he is "himself a present protagonist, who 'speaks' [...his] experiences and selves as they occur, in a series of lyric comments which slowly add up to something like a 'contour of life'".[33] The speaker "is a present 'I' who recollects the links in the 'chain of experience' between himself...and the protagonist, the past 'I'".[34] This appears to match Hallam's sense of 'soul' as constructed by successive selves. However, modifying Gates's observations, the issue for the uniformitarian 'I' is precisely that the speaker's "experiences and selves" fail to "add up to" anything like a "contour of life". The self is given in the fragmentary workings of consciousness in the present and as such it is an open-ended 'I': a consciousness in process that cannot know itself as whole. Just as the speaker's past, present and future selves fail to add up to a whole, finalised self, the fragmented lyrics of *In Memoriam* compromise closure and struggle to add up to a cohesive poem. These forms of expression enact the Lyellian loss of history and teleological purpose—as Lyell's geology is a non-progressive, ahistorical observation of present process, the poem's construction

---

[31] Michael Holquist, *Dialogism: Bakhtin and his World* (1990), 29.

[32] Secord, *Principles* (1997), xviii.

[33] Gates (1999), 516.

[34] Ibid., 516.

of self is one that is outside narrative flow; present only in present consciousness. Thus, while *In Memoriam* is often seen as expanding personal grief into a wider grief for the species in the light of geological discoveries, here Lyell's anti-teleological geology is scaled down to an intimate level that imagines the mind bereft of its narrative.

At times when doubt prevails the speaker's experience of consciousness in the present leaves him distinctly detached from the past. In this state, his identity is experienced as provisional, impermanent and changeable in a present where there is no solid ground for the construction of knowledge or identity. In section XXIV, for example, the theme is the speaker's perception of the past, and doubt is the first issue:

> And was the day of my delight
> As pure and perfect as I say?
> The very source and fount of Day
> Is dash'd with wandering isles of night.
>
> If all was good and fair we met,
> This earth had been the Paradise
> It never look'd to human eyes
> Since our first Sun arose and set.
>
> And is it that the haze of grief
> Makes former gladness loom so great?
> The lowness of the present state,
> That sets the past in this relief?
>
> Or that the past will always win
> A glory from its being far;
> And orb into the perfect star
> We saw not, when we moved therein?

Here, amid the Lyellian instability of "wondering isles", a displacement occurs as the past which "looms so great" is uplifted in unison with the depression of the present; the "lowness of the present state". Moreover, the speaker is a present speaker whose reflection on the past is meant to shore up the self's sense of identity. However, the past appears to have lost its authority over the present, and the speaker's questions suggest an anxiety over the reliability of memory and identity. He asserts two possibilities for the idealising effect of memory: either present grief makes the past stand out—"The lowness of the present state, /…sets the past in this relief". Or distance glosses the past—"the past will always win / A glory from its being far". In both cases, recognising how the present dissolves the fixity of the past leads to doubt concerning its authenticity. The third stanza, foregrounding displacement, contrasts the "lowness of the present state" with a "pure and perfect" past gesturing towards Adamic epistemologies that see in the present state of humankind a falling away from Adamic perfection. However, in the second stanza, the speaker radically revises his position by stating that the "Paradise" of the past has never been seen by "human eyes", a statement that appears to reject Adamic theory altogether. Doubt about the speaker's own perception of his personal past broadens into a wider doubt concerning the validity of Adamic and teleological ways of knowing. Again both micro and macro states of uncertainty, like that other micro/macro crisis registered in *In Memoriam* (the extension of individual death into species extinction), are geological, while, in this case, they are the result of experiencing the self via uniformitarianism's anti-teleological prioritising of the present over the authority of the past.

The speaker's personal sense of self is destabilised by his distrust of the past. Not to trust the past indicates an anti-Romantic rejection of the primacy of experience and of the Wordsworthian sense of self as sustained by memory. The speaker looks back to the past from the 'I' experiencing itself in the present and asks "was the day of my delight / As pure and perfect as I say?" Here, the emphasis is not only on the uncertainty of the past but on the unreliability of the present 'I' to make judgements about it. The present speaker addresses his present ability to perceive—the question is not "was the past really perfect?" but "is my present perception of the past reliable?" This compounds the problem which is not only one of doubting the past but also one of self-doubt, as the past is now seen as essentially unknowable, arbitrarily and subjectively constructed in the present. More troubling, however, is the implication for selfhood, as, if memory (which ostensibly ensures selfhood) is unreliable, even false, if all that goes to make up identity is consciousness in the present, then what is it anyway that can transcend the material self? What can possibly transcend the physical self if consciousness and identity itself is merely a facet of an ever-changing present? Such a way of being makes Hallam's theorised 'one subject' unattainable, just as in Lyell's geology, Whewell's "one cycle" narrative of creation is lost in repetitive, non-progressive process. The past becomes, for the speaker, the arbitrary production of the present. He is left in an alien and humanly inhospitable landscape, expelled from the idealised landscape of pastoral memory and cast into one where natural laws and human memory are no longer in sympathy.

In the fourth stanza, the alternative view of how the past is idealised results in a similar conclusion. The collective perception of the "glory" that "We" see in the past, once we are distanced from it, is merely a smoothing away of the imperfections of the planet. The past "orb[s] into the perfect star" to create something like that early geological vision of paradisal beauty—the perfectly round, smooth earth with "not a wrinkle, scar or fracture in all its body", that described the "Golden Age" envisaged by Thomas Burnet in his *Sacred Theory of the Erath* (1680–90).[35] The "perfect star" of the paradisal past is an Adamically figured past in which divine authority is exemplified. However, in the poem, that past is taken to be illusionary, merely the effect of distance. The speaker's uniformitarian premise re-visions the past from the position of present perceptions, so that, as in Lyell's geological methodology, the past is presumed to be no different from the present. Paradoxically, the "perfect star" of the past not only offers a view Burnet's geological paradise, it also offers a uniformitarian perspective on the earth in which geological features are diminished by time and distance. Time subsumes the greatest changes, and distance smooths away the inequalities of the landscape, so that the highest mountain, Lyell tells us in an attempt to offset his readers' anthropocentricism, "would only be represented on an artificial globe, of about six feet in diameter, by a grain of sand less than one-twentieth of an inch in thickness" (I, 113). 'Orbed' "into the perfect star", the past, which seemed to resemble Burnet's paradisal past, is in fact a uniformitarian vision of the earth in which the hierarchy of the past—the Adamic authority that lies behind divine creation—is levelled by uniformitarian deep time and its assumption of sameness. In this re-visioning, there can be no recourse to the authority of the past, no Arcadian memory and no pastoral landscape of elegiac recuperation.

The idealised pastoral landscape of the elegy is replaced by a Lyellian landscape, and on this alternative formal ground, where repetitive stanzas operate to immutable formal laws with "no vestige of a beginning, no prospect of an end", hope for recovery is continually foiled. In line with the poem's displacement, and matching the debilitating episodes of doubt, are rapturous expressions of love and moments of faith. At times,

---

[35] Thomas Burnet, *Sacred Theory of the Earth,* published originally in four volumes between 1680 and 1690 (1719), 90–1. Remarkably, Burnet's text saw its final edition as late as 1828, and significantly, no further editions appear after the publication of Lyell's *Principles* in 1830–1833.

the span of life is a "tract of time", and the five years with Hallam, "the richest field". Love too, in deference to desire, is also a solid, measurable and manageable landscape, the five remembered years a "province not large, / A bounded field, nor stretching far" (XLVI, 8–14). But these investments in an idealised landscape indicate the speaker's avoidance of the logical conclusion that by tying memory and love to landscape he leaves them both open to the continual, insidious effects of geological action; this is an avoidance that represents a lapse in what is otherwise a lucid realisation that memory and love are intangibles solely contingent on present consciousness. The "eternal landscape" of the pastoral elegy's idealised past—that "richest field" of the memory of love—is finally made unavailable under the uniformitarian gaze, where past and future are no different from the present. This is a failure of the function of the poem to perform recovery but a failure that emerges from the poem's success, in that this *is* an elegy and one that touches its readers with its honest doubt and its willingness to enter a darkness in which all form is potentially lost.

In seeking to remain present in the materiality of existence where Hallam's "loved remains" (IX, 3) still remain, the poem structures itself around Lyellian uniformitarian laws. The elegiac poetic plane which ostensibly facilitates recovery by assimilating material loss into its idealised landscape of the past is replaced by the Lyellian landscape of the perpetual present where, it is hoped, material remains remain in the present and cannot pass into memory. However, by forming itself on Lyell's laws, the poem must also abide by Lyell's strategy of division which insists on an inviolable division between the physical/material world and the world of human morality and intellect. Thus, in *In Memoriam*, sympathy between nature and soul is severed. The Lyellian landscape dissolves the elegiac ground from under the speaker's feet, turning the "lavish hills" (XXIII, 11) of Arcadian pastoral memory into the "shadow hills" (CXXIII) of an aimlessly shifting earth emptied out of meaning. To end the poem, the speaker must finally give up Hallam's physical remains and follow the metaphysical path to a transcendence in which all selves are lost in the "general soul"—a consolatory condition which cannot satisfy desire. The poem asks: where are Hallam's remains? Its answer is to show that in the Lyellian landscape they are condemned, like all remains, to be "sealed within the iron hills" or "blown about the desert dust" (LVI, 19–20). However, Hallam's remains are also in the "fall'n leaves which kept their green, / The noble letters of the dead" (XCV, 23–4)— his *Remains in Verse and Prose*—but this is to locate the self in text—in language. Textual remains supply the speaker with his nearest experience of reunion as, "word by word, and line by line, / The dead man touch'd me from the past" (33–4), but this reunion with a textual self only accentuates the lack of material presence—language remains only ever secondary and unsatisfactory.

In Memoriam's Teleological Narrative

For all its problems with closure, *In Memoriam* does, of course, end. Closure is forged from a thematic diversion away from the poem's material structure to generational development interwoven with notions of death as spiritual development: "That men may rise on stepping-stones / Of their dead selves to higher things" (I, 4). Chambers's *Vestiges* has frequently been linked with what has been seen as the poem's evolutionary language. *Vestiges* speculates on humanity's place in the progressive development of 'animated nature':

> Is our race but the initial of the grand crowning type? Are there yet to be a species superior to us in organisation, purer in feeling, more powerful in device and act, and who shall take a rule over us!…There may be occasion for a nobler type of humanity, which shall complete the zoological circle on this planet, and realize some of the dreams of the purest spirits of the present race.[36]

In Memoriam looks forward to the coming of the "crowning race" and Hallam is representative of "a noble type" (Epilogue, 136, 138). However, while Tennyson seemed to have absorbed the language of *Vestiges*, his configuration of spiritual progress is not necessarily evolutionary. Rather, the poem reverts to a teleological narrative taken in part from Hugh Miller's *The Old Red Sandstone* (1844). Miller's specific notion of geological and spiritual progress—so important to *The Princess*—finds its way into *In Memoriam*'s formulation of a geologically oriented progressive spiritual development. As Chap. 2 argued, Miller's notion of progress was not evolutionary, as Chambers's was, but progressive in terms of serial creations. While for Lyell, anything less than the possibility of full development in the present was untenable (see, for example, Lyell's anti-evolutionary

---

[36] Robert Chambers, *Vestiges* (1844), 267.

accommodation of the embryonic theory discussed in Chap. 3), Miller saw progressive creations as the pattern through which both God and nature move. For Miller, the fossil record showed no signs of evolutionary progress, but instead indicated the existence of complex life in the earliest epochs as well as the most recent strata of each epoch. However, each new epoch represented an advancement in the complexity of life.

Towards the end of *The Old Red Sandstone*, Miller draws his conclusions about fossil evidence into a divine grand narrative of teleological progress as a future epoch is envisaged as heralding a new and 'nobler' creation:

> Has the last scene in the series arisen, or has the Deity expended his infinitude of resource, and reached the ultimate stage of progression at which perfection can arrive? The philosopher...decided in the negative, for he was too intimately acquainted with the works of the Omnipotent Creator to think of limiting his power; and he could, therefore, anticipate a coming period, in which man would have to resign his post of honour to some nobler and wiser creature,—the monarch of a much better and happier world. (*ORS*, 274)

Miller ends his geological treatise with an ecstatic endorsement of spiritual progress:

> How well it is...to be able to look forward to the coming of a new heaven and a new earth, not in terror, but in hope,—to be encouraged to believe in a system of unending progression, but to entertain no fear of the degradation or disposition of man! The adorable Monarch of the future, with all its unsummed perfection, has already passed into the heavens, flesh of our flesh, bone of our bone, and Enoch and Elias are there with him,—fit representatives of that dominant race, which no other race shall ever supplant or succeed, and to whose onward and upward march the deep echoes of eternity shall never cease to respond. (*ORS*, 274–275)

As Miller ends, *In Memoriam* also ends with a similar endorsement of the "one far-off divine event, / To which the whole creation moves" (Epilogue, 143–144). Adopting Miller's geology of progressive creation allows a thematic transcendence of the rigid structure of the poem's form. The language of progress is initiated by the dream of section CIII, which, introduced towards the end of the poem, stages an alternative narrative of linearity and progress designed to counter Lyell's fixed laws of continual non-progressive displacement. The licence of the dream sanctions a cessation of Lyell's ostensibly rational laws, and offers to repair the Lyellian division between the human and the natural worlds. The dream's progressive language offers a model of spiritual and moral growth compatible with teleological epistemologies. What the speaker requires that uniformitarian unending process could not supply is an anthropocentric function for nature's laws and an eschatological purpose for Hallam's death. The dream provides both, envisaging Hallam as a Christ-like figure who leads humanity toward a higher stage of development, from the physical "first Form" (LXI, 10) to the spiritual "second state sublime" (LXI, 10).

Miller's language of "upward and onward" progress initiates an ascending spiritual journey which begins in the material: the "hall" subject to the geological agents of aqueous erosion in the "river sliding by" (5, 8), but increasingly moves upward into a nebulous immateriality. The tides are "forward-creeping" (37), the approach to Hallam's "great ship" (40) is "Up the side" (43), and the speaker and his companions move "toward a crimson cloud" (55). In keeping with the aim of abolishing Lyell's strategy of division, the dream's language of progress conflates physical and spiritual growth. The journey of the speaker and the maidens towards Hallam's "ship" reveals a "vaster" (25) shore and "grander space" (26). The spiritually developed dead Hallam is "thrice as large as man" (42). The speaker feels within himself "The pulses of a Titan's heart" (32), and "the thews of Anakim" (31), the latter reference alluding to the Biblical tribe of *Deuteronomy* I:28, whose "people is greater and taller" and whose "cities are great and walled up to heaven".[37] Similarly, as the journey continues, the maiden muses "gather'd strength" and "presence, lordlier than before" (27). Spiritual growth has a physical dimension that not only helps to imagine that longed-for union with Hallam the physical man, but also invests physical process with spiritual meaning to institute a profound morphological transmutation in which both the material and the physical self are essential ingredients.

Fittingly, in the dream it is Hallam's physical representation that is the centre of worship:

---

[37] It should be noted that there is a formal as well as a metaphoric interaction between section CIII and *Deuteronomy*. Alan Sinfield points out that *In Memoriam*'s "syntax is distinctly reminiscent of the Bible" by making a comparison between the seventh and eight stanzas of section CIII and *Deuteronomy*. See Sinfield (1971), 91.

The hall with harp and carol rang.
They sang of what is wise and good
And graceful. In the centre stood
A statue veil'd, to which they sang.

The "statue veil'd" represents Hallam's remains, and while his remains
are left behind in the "flood below" (CIII, 20) to the ultimate fate of the
planet, like Miller's Christ who is the "adorable Monarch of the future",
Hallam has "already passed into the heavens, flesh of our flesh, bone of
our bone" to become one of the "fit representatives of that race, which
no other race shall ever supplant or succeed". Like the statue of Sir
Ralph that represented the "master existence" of the past in *The Princess*,
Hallam's "statue veil'd" is the type of a future "master existence" veiled
and as yet unknowable. The dream counters nature's 'evil dream' that
God and "Nature" are "at strife" by offering in its place a dream of physical and spiritual union, coupled with a narrative of progress. Working
in opposition to the poem's material non-progressive Lyellian form, the
dream, we are told, leaves the speaker "content" (CIII 4). It introduces
the poem's teleologically figured Epilogue, which reinstates human narratives with its vision of birth and death and its figuration of first and
final causes. The goal at the end of the poem is to envisage a complete
knowledge of nature, an assimilation of nature into human consciousness that offers in opposition to Lyell's strategy of division the ultimate
unity of the physical/material and spiritual/moral. This one knowledge,
like Whewell's consilience, makes all things subject to it; under its "com-
mand / Is Earth and Earth's" and, for those who wield this knowledge,
"Nature [is] like an open book" (130–2).

The final lyrics of progress have been read as evolutionary:

No longer half-akin to brute
For all we thought and loved and did,
And hoped, and suffer'd, is but seed
Of what in them is flower and fruit. (133–6)

These lines have been read as presenting a paradox: "Even as Tennyson
dreams of the break from the biological, his metaphor is drawn from
botanical sexual reproduction".[38] But the metaphor is fitting, as what

---

[38] Tomko (2004), 130.

at first appears as a contradictory use of biological imagery to figure a process in which the physical self is outgrown, is, in fact, a deliberate attempt to heal over Lyell's division through the figuration of spiritual progress as physical process—the spiritual, biological, human and geological function that Miller saw as a "system of unending progression" towards "eternity". The 'truth' of spiritual progress finds its evidence in geology but not in Lyellian geology, which the poem cannot reconcile with progress, but in the "master existences" that Miller's geology suggests evince divine progress in serial creation—the "master existences" that Hallam's remains, memorialised in the "statue veil'd", exemplify.

Despite the language of spiritual progress, the text's most immediate desire for union with the dead Hallam must remain unfulfilled. The dream is only a temporary suspension of the 'rationality' imposed by *Principles*. *In Memoriam* cannot escape its non-progressive Lyellian form, which so perfectly epitomised unending grief, and, as already noted, Tennyson felt that *In Memoriam*'s ending was too optimistic, indicating that he had not yet finished his uniformitarian poetic experimentation. Tennyson, in some senses, picked up where he left off when he began *Maud*. Here, the poem's dramatic persona offers some distance from the rawness of the uniformitarian present, leaving Tennyson freer to take Lyell's geology to its logical conclusion. The speaker of *Maud*, of course, is at points in the poem mad, and his madness is the result of his full engagement with the dialogic potential of uniformitarianism. This self is the subject of Chap. 6. Before that, however, Chap. 5 examines *Maud*'s speaker in his interaction with remains (organic and inorganic, geological and memorial) and the provocative narrative of self that these remains create.

CHAPTER 5

Reading *Maud*'s Remains: Geological Processes and Palaeontological Reconstructions

> *To what base uses we may return, Horatio!*
> *Why may not imagination trace the noble dust*
> *Of Alexander, till he find it stopping a bung-hole?*
> (*Hamlet* 5.1. 202–4)

As Tennyson's "little *Hamlet*", *Maud* (1855) posits a speaker who, like Hamlet, confronts the ignominious fate of dead remains.[1] *Maud*'s speaker contemplates such remains as bone, hair and shell, and he experiences his world as one composed of hard inorganic matter—such things as rocks, gems, flint, stone, coal and gold. While *Maud*'s imagery of "stones, and hard substances" has been read as signifying the speaker's desire "unnaturally to harden himself into insensibility", this chapter argues that these substances benefit from being read in the context of Tennyson's wider understanding of geological processes.[2] Along with highlighting these materials, the text's imagery focuses on processes of fossilisation, with *Maud*'s characters appearing to be in the grip of an

---

[1] Tennyson called *Maud* "a little *Hamlet*." See *Memoir*, 1: 396.

[2] John Killham, "Tennyson's *Maud*. The Function of the Imagery" in *Critical Essays on the Poetry of Tennyson*, John Killham ed., (1964), 231, 235. J.L. Kendall has also explored the ambiguity of gem imagery, asserting that the speaker, "through the intermediate symbol of stone … vaguely or unconsciously associates gems with death". See, "Gem Imagery in Tennyson's *Maud*" (1979), 392.

© The Author(s) 2017
M. Geric, *Tennyson and Geology*, Palgrave Studies in Literature, Science and Medicine, DOI 10.1007/978-3-319-66110-0_5

insidious petrifaction. Despite the preoccupation with geological materials and processes, the poem has received little critical attention in these terms. Dennis R. Dean, for example, does not detect a geological register in the poem, arguing that by the time Tennyson began to write *Maud*, he was "relatively at ease with the geological world".[3] This chapter argues, however, that *Maud* reveals that Tennyson was anything but "at ease" with geology. Where *The Princess* had confidently conflated the fossil bones of the "vast bulk" and Sir Ralph's stone statue to stand as evidence of the naturalness of "master existences" and the aberrance of feminism, in *Maud*, past master existences are figured as alien and hostile. Here they are recast as the "monstrous eft" that was once "Lord and Master of Earth", or, in reference to more recent remains, the even more grotesque "Assyrian Bull" of some enigmatic, semi-bestial race. By the time Tennyson got to *Maud*, Miller's rhetorical flourish seems to have lost its persuasiveness. Where *In Memoriam* wrestled with doubt initiated by geological theories, finally rising above Lyell's uniformitarian landscape in the onward and upward development of creation, *Maud*, conversely, seems to gravitate towards the ground, concerning itself with the corporeal remains of life and with the agents of change that operate on all matter. *Maud* appears much more concerned with the processes of fossilisation as described particularly in Lyell's provocative writings on the embedding and fossilisation of organic material in strata in *Principles of Geology* volume 2. Tennyson also brings to the poem his own experience, as by the time he began to write *Maud* he had his own field research to draw on for his imagery, and much more hands-on experience of geological remains. He sought out such evidence in his travels, and, as a result, *Maud* feels more physically and viscerally engaged with geological remains. Thus, quite remarkably, *Maud* not only encounters the remnants of death in remains—confronting and reading their import—it goes further to probe the taphonomic processes that result in the incorporation of dead remains and even living flesh into the geological system.

This chapter begins with an examination of a number of sources for Tennyson's geological and palaeontological thinking, as while a few of these have already been suggested by other critics, important sources

---

[3] Dean (1985), 21.

have been missed—sources that allow for a comprehensive re-reading of the whole poem in terms of these disciplines. I use the term 'remains' (as suggested in Chap. 1) because while *Maud*'s tropology is largely geological, the text embraces a more general notion of remains that includes along with fossils the more recent remains of the dead. For example, the ring made from his mother's hair that Maud's brother wears, as well as such objects as the "rock" that represents the father that "fell with him when he fell" (I: 8), the empty shell of Part II, and the "jewel-print" of Maud's feet (I: 890), which, like a fossil imprint, offers the speaker the desired object via the contemplation of its conspicuous absence.[4] The term 'remains' can include palaeontological objects, as well as other relics, and it fits well with William Whewell's use of the term *palaeontology* in the 1840s. Whewell employed palaeontology, as Martin Rudwick has observed, "in an unusual, broad, and literal sense, to denote the study not just of fossils but of *all* entities of which the relics survive from the unobservable deep past".[5] 'Remains' in *Maud* represent any readable relics or fragments that allow the speaker to reconstruct the past as a narrative in the present, a narrative that in turn constructs the poem. In this attention to remains, Tennyson was experimenting with the "powerful new methods of comparative anatomy" that had proved so innovative in the recovery of remote and alien forms of life from fossil fragments.[6] Provocatively, Tennyson's friend Richard Owen (the most celebrated British comparative anatomist of his time) asserted that through the application of the principles of comparative anatomy, practitioners "have been enabled to restore and reconstruct ... species that have been blotted out of the book of life".[7] At times, the speaker's gaze is similar to that of the palaeontologist who reinvests remains with meaning in an attempt to "restore" the dead and write them back into the "book of life". *Maud* takes up the challenge of reconstruction. Gideon Mantell, as discussed in Chap. 1, invested the comparative anatomist with the ability to "reassemble" what

---

[4] All references to Tennyson's *Maud* are taken from *Tennyson's Maud: The definitive Edition*. Susan Shatto ed. (1986).

[5] Martin J.S. Rudwick, *Worlds Before Adam* (2008), 550.

[6] Martin J.S. Rudwick, *The Meaning of Fossils* (1985), 107.

[7] Richard Owen, *Lectures on the Comparative Anatomy and Physiology of the Vertebrate Animals*, 2 vols. (1846), II, 3.

would appear "to a person uninstructed in the science" as "a confused medley of bones and of osseous fragments".[8] In *The Princess*, the medley (the confusion of archaeological and geological remains) is harmonised by Miller's "one act" of creation. In *Maud*, however, such cohesion is not available to a speaker who finds himself expelled from the ideological centre and increasingly aware that the meaning of remains is contingent on the observer's subjectivity. In this highly intimate poem (see, for example, Ralph Rader's biographical reading of 1963), Tennyson gets close up and personal, evoking remains in the service of his speaker's reconstructions of a highly subjective past. Examining this use of remains, the second section of the chapter focuses on the speaker's construction of the narrative of the past, offering close readings of the speaker's own readings and reconstructions of the objects he confronts.

Tennyson's imagination seems to have been energised by the extraordinary claims made concerning the ability of the new science of comparative anatomy to reconstruct a material past. Analogies between 'reading' fossil fragments and Emerson's notion of language as 'fossil poetry' (an analogy taken up by Richard Chenevix Trench, as discussed in Chap. 1) showed the close proximity of geology and language theory. The idea of writing the dead back into the "book of life" draws attention to the textual nature of the process of reconstructing the past and to how existing narratives provide contexts for remains without which they are merely "a confused medley" of unreadable fragments. The task of rewriting the dead back into the "book of life" thus entails assimilating remains into contemporary taxonomic and teleological contexts—writing them, in other words, into the meaningful narratives of the present. Virginia Zimmerman points out how geologists, palaeontologists and archaeologists "fashioned narratives out of fragmented remains", and notes that their "authority, rooted in [...the] ability to read well, lends similar authority to any reader—with the ability to interpret comes narrative authority".[9] In *Maud*, remains provide a means to (re)write the past in the contexts of the present. The speaker reads remains as a way of rewriting the fragmented remnants of a past that presents itself to him as a "confused medley" of events. Remains

---

[8] Gideon Mantell, *The Wonders of Geology Or, A Familiar Exposition of Geological Phenomena*, (1838), I, 127. Dean suggests the influence of Mantell's geological text on *In Memoriam* (10–11).

[9] Zimmerman, *Excavating Victorians* (2008), 2, 38.

allow him not only to rewrite the past into a narrative that is sympathetic with his perception of events, but also to claim authorship and therefore authority over that past. The speaker's compulsion to reconstruct remains (as expressed, for example, in his attempt to reconstruct the "little living will" (II: 62) of the shell) is the compulsion towards narrative and towards the dynamic production of a textualised self that is spatially and temporally posited in the continuum of narrative flow. Thus, for the speaker of the poem, reading remains is not only a way of writing the object back into being but also a way of writing the self into being.[10]

At points, however, remains serve a different purpose for the poem's rhetoric. They are used to emphasise how all living things, and even entire civilisations, inevitably become the remains for a future age to contemplate. And *Maud* articulates specific concerns not only for how the past is read from its remains but also for how future generations will read the remains of the present age—an anxiety implicit in the text's disturbing figuration of processes of petrifaction and fossilisation occurring in the present. Just as geological catastrophe had arrested Pompeii at a moment in time that laid bare for future generations its less heroic quotidian concerns, petrifaction in *Maud* encodes an anxiety about how contemporary remains, similarly arrested in time, might be read by future generations and what these remains may say about the present Mammon-worshiping world.

As Zimmerman suggests, the contemplation of remains "forces the observer to redefine himself in relation to [...an] ever-expanding time scale and to imagine his own end as a similar artefact".[11] The speaker of *Maud* imagines not only how the present age will be read in the future but also how his own organic remains might be read, or, more troublingly, how they might not be read—how they might, in fact, be "blotted out of the book of life". Such fears stem not only from concerns about the type of readable signs that remains leave behind them,

---

[10]This reading is indebted to E. Warwick Slinn's brilliantly detailed analysis of the Victorian long poem, and specifically *Maud*, in the context of Hegelian dialectics. See Slinn, *The Discourse of Self in Victorian Poetry* (1991), particularly Chap. 2, "Consciousness as Writing" and Chap. 3, "Absence and Desire in *Maud*". More broadly, it is also indebted to Isobel Armstrong's ground-breaking reading of the Victorian double poem, in *Victorian Poetry: Poetry, Poetics and Politics* (1993).

[11]Zimmerman, *Excavating Victorians* (2008), 14.

but also from the way the narratives constructed from remains raise questions about interpretation itself, as such narratives are the constructions of the present and are therefore arbitrary and unreliable indicators of the past. They are, in fact, merely the present observer's projection of meaning over the non-existent, empty space of the past. The methods used in comparative anatomy typify this problem, as fossil fragments can only be understood within an acknowledged system: it is an episteme that rests on the premise that the reconstruction of organised bodies from fossil remains is possible through reference to organised bodies in the present. Reconstructing remains, as already suggested, is, in fact, the act of constructing the past within the taxonomic narratives of the present.

Remains are in themselves tangible memorials of the missing self. As John M. Ulrich puts it: "material remains are the provocative remnants of a past once vibrant and living, but in and of themselves, such remains are just that remains—partial, dead, silent, and other".[12] For *Maud*'s speaker, for whom remains become the materials of narrative recovery, the problem he confronts is that such remains signify in themselves silence and death and the narratives they construct only compound his sense of himself as dissociated from a real and living world—as "nameless" (I: 119) and as already dead. Thus, where *In Memoriam*'s geology effectively expanded individual grief into a grief for the collective extinction of the species, *Maud*'s geology expands the speaker's personal crisis into a wider existential experience of meaninglessness. The speaker's psychological crisis finds expression through the confrontation with geological processes and images, and while Lyell's writings are largely responsible for initiating such a crisis, they also offer a solution of kinds to the problem of recovering the past, as discussed at the end of the chapter—a solution at least to the speaker's specific fear that his own remains might not leave a readable trace.

SOURCES

Maud's preoccupation with remains was the result of a conflation of themes and images that Tennyson explored in the months leading up to his writing of the poem. Geology was never far from Tennyson's

---

[12] John M. Ulrich, "Thomas Carlyle, Richard Owen, and the Paleontological Articulation of the Past" (2006), 45.

thoughts and in November 1853 (a year before he sat down to write *Maud*), the family moved to Farringford on the Isle of Wight, where Tennyson furthered his interest in geology and "trudged out with the local geologist, Keeping, on many a long expedition" (*Memoirs*, 1: 366). On May 19, 1854, on a trip to the mainland, Tennyson pre-viewed the Crystal Palace Exhibition at Sydenham, just weeks before the official opening, writing to his wife that it was "certainly a marvellous place" (*Letters ALT*, II, 90). The spectacular potential of the emerging sciences of comparative anatomy and archaeology to recover the past were powerfully represented in the exhibition, and of all its wonders, Tennyson notes that he was "much pleased with the Pompeian house and with the Iguanodons and Ic[h]thyosaurs" (*Letters ALT*, II, 90).[13] The latter were the first life-size reconstructions of prehistoric animals ever to be exhibited. They were created by the sculptor Benjamin Waterhouse Hawkins under the direction of Richard Owen.[14] Tennyson had met Owen for the first time some two years previously (according to Owen's diary) on August 6, 1852, and Owen was to become Tennyson's companion on geological field trips in later years on the Isle of Wight.[15] The iguanodon and ichthyosaurs were placed, along with reconstructions of other extinct animals, on "Islands in the Geological Lake", which had been constructed to represent five geological epochs, complete with corresponding rock formations and strata. As Owen explained in the exhibition guide, the purpose was to "demonstrate the order of succession, or superposition, of these layers or strata, and to exhibit, restored in form and bulk, as when they lived, the most remarkable and characteristic of the extinct animals and plants of each stratum".[16] The reconstructions provided a fabulously exotic spectacle that must have thrilled Tennyson's geologically orientated imagination and that offered him a remarkable confrontation with the power of comparative anatomy to, as Owen had put it, "restore and reconstruct ... species that been blotted out of the book of life". As Paul Turner notes, Owen's prehistoric figures reappear

---

[13] Also see *Memoir*, 1: 376, which gives the date of the visit as May 22.

[14] For a description of the process of restoration, see Richard Owen, *Geology and Inhabitants of the Ancient World* (1854).

[15] Richard Owen, *The Life of Richard Owen* 2 vols. (1894), 1, 388–89.

[16] Owen, *Geology* (1854), 7.

as *Maud*'s "monstrous eft" who was "of old the Lord and Master of Earth".[17]

The reconstructions, Owen admitted, might "by some ... be thought, perhaps, too bold". However, they were justified, he claimed, by the way they demonstrated the "successive periods, during which many races of animated beings, distinct both from those of other periods and from those now living, have successively peopled the land and the waters".[18] The exhibition thus illustrated and emphasised the "discovery of the law of succession of animal life on this planet"; a law that Owen argued comparative anatomy proved.[19] The vista of succession did not end, however, with the animal reconstructions. Inside the Crystal Palace, ten stunning architectural and historical "'restorations' of buried empires" were on show in the successive Courts. The "Courts were to be a main feature of 'the education of the eye', to form a three-dimensional and full-colour encyclopaedia of the 'complete history of civilisation'".[20] Thus, they continued the depiction of successive "masters" of the earth in reconstructions of the art and architecture of past cultures, and in turn, they helped to weave geological history into more recent human history in the minds of visitors. The exhibition emphasised a continual succession of animal types and human civilisations in the way it left unrecorded the aeons of geological time that divided Hawkins's dinosaurs from each other and from the world of the modern viewer. It collapsed time, visually enforcing Owen's "law of succession" in which an unvarying and regular law of the rise and fall of "races of animate beings" works through the whole of Earth's history.[21]

---

[17] Paul Turner, *Tennyson: Routledge Author Guides* (1976), 136. Turner's essay provides a brief but highly valuable account of the possible sources and influences informing *Maud*.

[18] Owen, *Geology* (1854), 7.

[19] Owen, *Lectures* (1846), II, 3. An examination of the ideological thrust behind Owen's science is beyond the scope of the present study. However, Rudwick suggests that the arrangement of extinct animals in terms of successive "races of animated beings" illustrates Owen's determination to read fossils as "authoritative evidence against the rising tide of evolutionary speculation in the Lamarckian mode" (*Scenes* 142). Also see Rupke, who writes, "Any antitransformist applications of Owen's reptilian reports were, admittedly, a welcome side effect, part of Owen's overall Cuvierian mission; [although] not a formative concern" *Richard Owen* (1994), 81.

[20] Jan Piggott, *Palace of the People: The Crystal Palace at Sydenham 1854–1939* (2004), 67.

[21] Owen, *Geology* (1854), 7.

Owen's vision of continual succession, however, was a troublesome one, as while the Crystal Palace proclaimed the mid-Victorians' consummate ability to read the past from its remains and thus their assumed mastery over the past, it also suggested a law at work in the nature of things from which the they could not extricate themselves. In this way, the exhibition encouraged Victorians to extrapolate Owen's vision of succession and to envisage their own inevitable extinction. Zimmerman demonstrates how the Victorians were acutely aware of the fate of their own civilisation as the potential ancient relic of the future, as extensive building work across the capital, in what Zimmerman calls "accidental archaeology", increasingly revealed London's buried Roman past. Such discoveries demonstrated how the land is peopled successively, since not only was London once the scene of Roman otherness, but the monsters of an inconceivably distant past had also once roamed the land on which London now stood. Nancy Rose Marshall also points out that the Palace's journey in time was in reverse: "Moving away from the displays of present civilisation and human history in the Palace, the visitor travelled further back in time, as he or she crossed the grounds, a temporal regression characterized as a movement into the wilderness." Thus, she argues, "the park's backward spatial model of time" encouraged "thoughts about human extinction".[22] The Crystal Palace also emphasised how access to history in geological, palaeontological and archaeological terms relied upon excavating and reading the fragments and relics of a buried past, as all things successively gravitate towards the ground and are embedded in the earth's strata to become the matter upon which future worlds are built.

Inside the Crystal Palace, in the Nineveh Court, Tennyson would have seen reproductions of the winged Assyrian bulls discovered by A.H. Layard, which, as critics have noted, gave him his description of Maud's brother—"That oil'd and curl'd Assyrian Bull" (I: 233).[23] Tennyson read Layard's *Nineveh and its Remains* (1849) in the summer of 1852 (*Memoir* 1: 356). The Nineveh Court had a particular resonance for Victorians, as Assyria was "associated in contemporary minds" with the

---

[22] Nancy Rose Marshall, "'A Dim World, Where Monsters Dwell': The Spatial Time of the Sydenham Crystal Palace Dinosaur Park" (2007) 289, 296. Marshall offers an interesting analysis of the contradictory ideologies of progress and degeneration embodied in the arrangement of the Crystal Palace.

[23] See Turner, *Tennyson* (1976), 145. Also see Shatto, *Tennyson's Maud* (1986), 181.

"biblical books of *Kings* and *Chronicles*, and the prophecies of Isaiah and Jeremiah about desolation".[24] The Nineveh Court "was a monument to imperial power and pride" and its remains were seen as a "prophetic *memento mori* of the possible demise of the British Empire". Similarly, the Pompeian Court, which particularly pleased Tennyson and which had the added drama of the city's tragic fall to geological catastrophe, was presented as a warning against decadent materialism. The Courts generally "suggested a certain politics of empire, a philosophy and even a morality: the fall of proud, wealthy and luxurious civilisations".[25] Visitors, many of who would have been familiar with Edward Bulwer-Lytton's hugely popular *The Last Days of Pompeii* (1836), were "told that Pompeii was the fashionable resort of a hedonistic class" and that "it would be helpful to think of it as the 'Worthing of Italy'".[26] The fate of Pompeii, in which organic material and even living individuals were apparently turned to stone, seemed a fitting Nemesis for a sensual and luxurious age. Rather erroneously, visitors were told, "Nineveh for all its pride fell in a day and it took only an hour for the sybaritic Pompeii to be buried in ash."[27] Thus, the grindingly slow progress towards civilisation was juxtaposed against the astonishing rapidity with which such progress is apparently undone.

Maud clearly draws on the rhetoric and narratives that the Crystal Palace implied, depicting a terminally materialistic age heading towards a manifestly geological fate. In the absence of Vesuvius, *Maud*'s nineteenth-century geological catastrophe involves an acceleration of geological time through which the text envisages petrifaction as occurring in the present. The age is presumably already morally dead—its soft parts, the human heart, for example, having been turned to stone. The speaker predicts early in the poem that "Sooner or later" he "too may passively take the print/Of the golden age", which will make his "heart as a millstone" and "set" his "face as a flint" (I: 29–31). He is "Gorgonised ... from head to foot" by Maud's brother's "stony British stare" (I: 464–65): his heart is "half-turn'd to stone", and is repeatedly

---

[24] Piggott, *Palace of the People* (2004), 75. Shatto points out the allusions to Job and Isaiah in *Maud*, 168.

[25] Ibid., 75.

[26] Ibid., 100.

[27] Ibid., 75.

described as made of stone (I: 267, 268; II: 132, 136). The unnatural speed with which fossilisation occurs parallels the increasing "lust for gold" (III: 39) which "gorgonises" and deadens all the characters of the poem in its valuing of dead geological materials such as coal and gold above the living individual. *Maud*'s characters appear to be already dead, and if not already petrified, they either seem to be in danger of slipping into the system of dead remains, or else they are represented by geological matter. Maud's brother is a "flint" (I: 740), a "lump of earth" (I: 537), and just as the speaker's father is represented by the "rock that fell with him when he fell", so Maud's mother, "mute in her grave", is signified by "her image in marble above" (I: 158). Maud herself, who is described as "dead perfection" (I: 83), is consistently equated with inorganic substances, being "gemlike" (I: 95), a "jewel" (I: 352), a "precious stone" (I: 498). Thus, she already belongs to a world of geological substances; her worth is measured not in terms of human values but geologically and economically. If Maud is represented by those geological materials that are most rare and coveted, then the poor are represented by more mundane matter. Fed with bread adulterated with "chalk and alum and plaster" (I: 39), the poor appear to undergo a grotesque transmogrification as they are slowly mineralised in a macabre transformation of living flesh into geological material. Thus, *Maud* depicts a fittingly geological destruction for the "Wretchedest age since time began" (II: 259). The ruin of the present world accedes to the inevitability of Owen's law of succession: *Maud*'s world is already solidifying, fixing its character into remains that will be read by the proficient reader (by a new "master existence" or "Lord and Master of the Earth") as the telling remnants of a materially oriented and morally defunct age.

The Crystal Palace exhibition visibly confirmed what Tennyson already knew from his reading of Lyell's *Principles*: that all life and all human artefacts are inhumed within the folds of the earth's strata as it relentlessly shifts through geological time. Lyell offered provocative visions of petrifaction and fossilisation, sometimes extrapolating the effects of geological processes in the present to visualise a distant future in which the present world is reduced to inorganic material. For example, contemplating deep time, Lyell writes:

> Let us suppose that at some future time the Mediterranean should form a gulf of a great ocean, and that the tidal current should encroach on the shores of Campania, as it now advances on the eastern coast of England:

the geologist will then behold the towns already buried, and many more which will inevitably be entombed hereafter, laid open in the steep cliffs, where he will discover streets superimposed above each other, with thick intervening strata of tuff or lava. (*PG*, I, 359–60)

The streets become strata in a passage that suggests how successive civilisations become subsumed into the earth while new manifestations of civilisation succeed them in a continual layering. Lyell goes on to suggest that, "Among the ruins will be seen skeletons of men, and impressions of the human form stamped in solid rock of tuff" (*PG*, I, 360). Mantell, in his geological writings, goes further, conjuring a distant future in which towns and cities will be constructed from materials that contain traces of the present human world.

The occurrence of human skeletons in modern limestone ... incontestably prove[s] that enduring memorials of the present state of animated nature will be transmitted to future ages. When the beds of the existing sea shall be elevated above the waters, and covered with woods and forest—when the deltas of our rivers shall be converted into fertile tracts, and become the sites of towns and cities—we cannot doubt that in the materials extracted for their edifices, the then existing races of mankind will discover indelible records of the physical history of our times, long after all traces of those stupendous works, upon which we vainly attempt to confer immortality, shall have disappeared.[28]

Here, organic traces of the age are incorporated into the geological system which in turn becomes the building materials that will raise new cities in an incomprehensibly distant future. Thus, it is the dead remains of the Victorians themselves that endure through time, not the great works of architecture or engineering designed with posterity in mind.

Even more disturbing, perhaps, was Thomas Carlyle's vision of remains and the readable history they leave behind. Carlylean "moral and social concerns" can be traced throughout *Maud*, as Michael Timko has shown.[29] However, the confluence of their thinking in terms of geological processes has not been noted. Like Tennyson, Carlyle was also influenced by Owen's comparative anatomy and Lyell's geology, as John

---

[28] Mantell, *Wonders* (1838), I, 114–115.
[29] Michael Timko, *Carlyle and Tennyson* (1987), 61.

Ulrich has strikingly demonstrated. Carlyle's "Bog of Lindsey", which provocatively evoked the successive layering of history as strata, was posthumously published, but, as Ulrich suggests, was "most likely [... written] in the latter half of 1843".[30] This was a time when Tennyson "was in the habit of walking out with Carlyle at night" and when they had "long and free discussions on every conceivable subject" (*Memoirs* I, 267).[31] In the "Bog of Lindsey", Carlyle writes:

> the leafy, blossoming, high-towering past century ... becomes but a stratum of peat ... the brightest century the world ever saw will sink in this fashion; and thou and I, and the longest-skirted potentates of the Earth,— our memories and sovereignties, and all our garnitures and businesses, will one day be dug up quite indistinguishable, and dried peaceably as a scantling of cheap fuel.[32]

Carlyle envisages the "great diversity of the past [as] compressed and homogenized" so that nothing distinguishable remains; all life is reduced to its lowest carbon common denominator, becoming merely fodder for the fire.[33]

Lyell writes in detail in the second volume of *Principles* on processes of fossilisation, discussing, for example, the various sites where ancient human remains had been found in the early stages of fossilisation. Tennyson seems to have been interested in observing such phenomena at first hand and keen to make literary use of his observations. Susan Shatto notes that in August 1854, shortly after his visit to Sydenham, he journeyed to Glastonbury, Wells and Cheddar "to gather materials and ideas

---

[30] Ulrich "Thomas Carlyle" (2006), 45. Ulrich also demonstrates that in March 1843 Carlyle "attended a series of lectures by Charles Lyell" and that "In a letter to Jane in July of that same year, Carlyle mentions reading 'Lyell's Geology'" (36).

[31] See also Thomas Herbert Warren, "FitzGerald, Carlyle, and Other Friends" in *Tennyson and his Friends*, Hallam T. Tennyson ed. (1911). Of Tennyson and Carlyle, Warren writes, "They foregathered a good deal at this period, [the early 1840s] sat and smoked silently, walked and talked together, both by day and night" (132).

[32] Thomas Carlyle, *Historical Sketches of Notable Persons and Events in the Reigns of James I. and Charles I*, 1898, Alexander Carlyle, ed. (1898), 64.

[33] Ulrich, "Thomas Carlyle" (2006), 47. Ulrich notes that at this time Carlyle was on friendly terms with Owen and was assimilating ideas from both comparative anatomy and geology (30). Thus, it is entirely conceivable that he and Tennyson discussed what were common interests.

... for his long-meditated Arthurian epic".[34] Here Tennyson explored a cave at Wookey Hole, noting, however, that it was "not quite what I wanted to see, tho' very grim" (*Memoir*, I, 377). What he expected to see seems to have come from his reading of Lyell, as Lyell specifically records how "human skeletons have ... been found in the cave of Wookey Hole ... dispersed through reddish mud and clay ... some of them united by stalagmite into a firm osseous breccias" (*PG*, II, 224). The "reddish mud and clay" of this cave was typical of sites in which human or animal remains are found in the process of fossilisation, and Lyell frequently notes the "reddish calcareous earth" (*PG*, II, 223), the "red osseous mud" (*PG*, II, 224), the "red breccias" or the "blood-red colour" (*PG*, II, 221) of strata at such sites. The cave, "tho' very grim", was, apparently, not quite as lurid as Tennyson had hoped. Returning to Farringford with the Crystal Palace and Wookey Hole fresh in his mind, Tennyson "worked at 'Maud' morning and evening, sitting in his hard high-backed wooden chair in his little room at the top of the house" (*Memoir*, I, 377).

From here, he wrote *Maud*'s arresting and emphatic opening lyric:

> I hate the dreadful hollow behind the little wood,
> Its lips in the field above are dabbled with blood-red heath,
> The red-ribb'd ledges drip with a silent horror of blood,
> And Echo there, whatever is ask'd her, answers 'Death.' (I: 1–4)

The "dreadful hollow", of course, is the site where the body of the speaker's father is found. Critics have commented on how the lines evoke a vividly sexualised landscape. As Jonathan Wordsworth noted in 1974; "once it has been pointed out it is difficult not to see the details of the first two lines ... in terms of the female body".[35] However, this resonant reading has tended to obscure the possibility of other interpretations, particularly in the light of Tennyson's reading of Lyell and his visit to Wookey Hole. The "red-ribb'd ledges [that] drip with a silent

---

[34] Shatto, *Tennyson's Maud* (1986), 17.

[35] Jonathan Wordsworth, "'What is it, that has been done?': The Central Problem of Maud" (1974), 358. Ricks describes the "surrealistic lunacy which suggests a menstruating woman" that the lines evoke, see, *Tennyson*, (1989), 239; while Linda Shires suggests that the lines express "disgust for a sexual female body", see, "*Maud*, Masculinity and Poetic Identity" (1987), 275.

horror of blood" encode the geological reality that the earth is essentially composed of the dead remains of organic life. They can be linked to the "reddish mud and clay" of the Wookey Hole cave, in which it was possible to find human bones. Moreover, on the Isle of Wight in the years after the publication of *Maud*, Tennyson explored what he called the "red cliff" for fossils with Richard Owen, writing afterwards to Owen, "we cannot afford to lose your brains ... not at least till all our lizards are dug out, and this stretch of red cliff which I see from the attic windows no longer needs interpretation" (*Letters ALT*, II, 406–8). Visible from the window of his attic study (the room in which Tennyson wrote much of *Maud*), the "red cliff", as Dean points out, is "an unusually interesting formation at Brook Bay of upturned ferruginous clays, sandstones, and shales capped by horizontal layers of gravel, clay and loam", and an area rich in dinosaur fossils.[36] Emily Tennyson also makes reference in her diary to the "red cliff" and "the wonderful dragon" that Owen made "out of the bones and scales" there found, suggesting that the "red cliff" was for the Tennysons the name given to this specific place of great palaeontological significance.[37] The "reddish mud" of the Wookey Hole cave and the fossil rich "red cliff" Tennyson viewed while he wrote *Maud*'s opening lines combined to create the image of the "red-ribb'd ledges". In this reading, the "dreadful hollow" is akin to the Wookey Hole cave and similar sites; it is not only a bloody site of death but also the place where the peculiar (and "very grim") processes of fossilisation occur, where an aggregate of bones and mud harden into rock through the pun in which the "red-ribb'd ledges" are indeed ridges of rib bones distributed through "reddish mud and clay".

The speaker's concern with remains begins with his anguished contemplation of his father's remains found in the "ghastly pit". Appropriately, the body fills the "pit" left by the "gutted mine" (I: 338) that was emptied out in the making of the "new made lord['s]" wealth. The fate of the body, which is "Mangled, and flatten'd, and crush'd, and dinted into the ground" (I: 7), evokes a type of fossilisation whereby, as Lyell explains, organic material is "squeezed down and flattened" in the making of coal.[38] Thus, the body appears to replenish the coal that

---

[36] Dean, (1985), 22.

[37] Ibid., 22.

[38] Charles Lyell, *Elements of Geology* (1838), 428.

is "all turn'd into gold" (I: 340) in a fitting exchange that exposes how the present age values dead remains above living individuals. Coal itself is the dead remains of organic life transformed in the process known as carbonisation—the early stages of which Carlyle alludes to in his vision of a future in which "thou and I", subsumed into the peat bog, are reduced to "cheap fuel". In carbonisation, as Lyell described it, "Sometimes only obscure or unintelligible impressions are left, and the lapidifying process has often effaced not only the characters by which the species, but even those whereby the class might be determined" (PG, I, 147–48). Remains are thus flattened, crushed and reduced to a carbon trace; a state in which taxonomic divisions become difficult or impossible to detect. The "characters", the readable signs that link past remains to the present (comparative anatomy's narratives of "class" and "genus"), are here effaced and therefore cannot be read. The "Mangled", "flatten'd, and crush'd, and dinted" body of the father is thus divested of social identity; the demarcations of "class" are squeezed out of his remains as he loses his footing in the social hierarchy and descends into the abyss of the "dreadful hollow". From here, the father cannot be restored and reconstructed, as his remains leave an unreadable trace—a circumstance acknowledged by the speaker, who expresses a concern that his father's good reputation will fail to be recorded: "his honest fame should at least by me be maintained" (I: 18). The decline of the speaker's own authority (his alienation from society, his namelessness and madness) point, however, to his inability to read, to rewrite, and therefore to recover the father within the wider shared social narratives of respectability and honour.

Carlyle, also connecting the mechanical actions of geological processes with the propensity of remains to lose their readability, writes of how "Generation under generation…and all higher generations press upon the lower, squeezing them ever thinner".[39] The squeezing and thinning out of remains and their increasing lack of distinguishability makes the recovery of the past impossible. The land in *Maud* has already been divested of meaning; the father's remains, which should be incorporated into "the dust of our heroic ancestors", fails to record a heroic history of the present for future generations to read.[40] Moreover, the land has

---

[39] Carlyle, *Historical* (1898), 64.

[40] Carlyle, *Latter-Day Pamphlets* (1850), 27.

been emptied out in the lust for coal and gold—a state of being that is the consequence of the greed and "Villainy" (I: 17) of Maud's father, and more generally of industrial capitalism represented by the new-made lord's grandfather, whose coal-fuelled fortune has been made in the creation of the "ghastly pit". It is this greed and villainy that led the speaker's father to the suicide's grave and thus to the "dreadful hollow" and the "ghastly pit" of anonymity where his battered remains, bereft of their honourable status, leave an unreadable and unrecoverable trace.[41] Thus, while the father's "honest fame" goes unrecorded, the petrifaction that overtakes *Maud*'s living characters leaves legible traces of an age deadened by, and insensible to, its own greed and villainy.

Lyell quotes Byron to evoke a sense of the land as comprised of dead remains: "The dust we tread upon was once alive!", further commenting:

> How faint an idea does this exclamation of the poet convey of the real wonders of nature! for here we discover proofs that the calcareous and siliceous dust of which hills are composed has not only been once alive, but almost every particle, albeit invisible to the naked eye, still retains the organic structure which, at periods of time incalculably remote, was impressed upon it by the powers of life.[42]

Owen makes an equally striking point in the Crystal Palace exhibition catalogue: "Chalk, immense as are the masses in which it has been deposited, owes its origin to living actions; every particle of it once circulated in the blood or vital juices of certain species of animals."[43] William Buckland too writes, "It must appear incredible to those who have not minutely attended to natural phenomena, that the microscopic examination of a mass of rude and lifeless limestone should often disclose the curious fact, that large proportions of its substance have once formed parts of living bodies." Buckland makes use of the sublime import of geology's findings to imagine how many of the "plains and massive mountains form ... the great charnel-houses of preceding generations, in which the petrified exuviæ of extinct races of animals and vegetables

---

[41] For an interesting examination of the meaning of the land in *Maud*, see Irene Basey Beesemyer, "'Black with the void from which God himself has disappeared': Spatial Displacement in Tennyson's Maud" (2000), 174–192.

[42] Lyell, *Elements* (1838), 56–57.

[43] Owen, *Geology* (1854), 9.

are piled into stupendous monuments of the operations of life and death."[44] He ends by quoting Cuvier who conjured in his writings the "imposing—even terrible—spectacle" that the dead remains of life form "almost all the ground on which we tread".[45] Not merely the receptacle of corporal remains, the earth is comprised entirely of dead remains processed into a state of indistinguishability under the grinding wheels of geological time. The "dreadful hollow", the "ghastly pit" and the "red-ribb'd ledges [that] drip with a silent horror of blood" all encode this macabre reality. They suggest not only that dead remains of life form "almost all the ground on which we tread" but also that these remains represent an infinitude of forgotten lives; the countless dead remains— from the speaker's dead father to the "little living will" (II: 62) that once inhabited the empty shell—all unceremoniously commingling, all essentially and irredeemably lost in myriad accumulations of dead remains over immeasurable time. In this sense, *Maud*'s remains record the levelling effect of geological processes and point to the ultimate breakdown of the distinctions of biological and social hierarchies. They remind the reader that, as Hamlet knew, the dust of the great may one day stop a "bung-hole".

More troubling, however, is that the body in the pit is "His who had given me life—Oh father! Oh God!" (I: 6), indicating not only the death of the earthly father but also the divine Father. Both are annihilated by geology; the physical father, whose geologically processed remains fail to leave a readable trace, and God the father, whose revealed truth is shown to be false by geological timescales and rendered absurd by the evidence of the rocks. Absence characterises both, as geology effectively empties them out leaving only the "dreadful hollow". This condition is indicated by "Echo" (I: 4), as "whatever is ask'd her" in this hollow place, her answer is "'Death'" (I: 4). Echo's voice foregrounds the deficiency of the speaker's personal narrative but also suggests a wider crisis specifically initiated by the geological enterprise itself. As a reflection of sound waves on hard surfaces, the echo is the enquiring geologist's voice directed towards the rocks and stones of a hard and uncomforting geological world. The answer of Echo, and the conclusion of geology, is that

---

[44]William Buckland, *Geology and Minerology Considered with Reference to Natural Theology*, 2 vols. (1836), I, 112–3.

[45]Rudwick, *Cuvier* (1997), 125.

death is the unequivocal law of nature; an answer that is all the more poignant as it represents the geologist's own voice reflected back to him. Thus, while the echo acting in a hollow represents the speaker's effort to overlay the "silent horror of blood" (I: 3) with the hum of human meaning, it only foregrounds a more profound sense of meaninglessness at the centre of being, one that can never be filled.

Reading Remains

Maud's speaker attempts to make sense of those past events that have resulted in his present circumstances. Like Hamlet, he is traumatised by the death of the father, dissociated from a world in which he has "hardly mixt" (I: 265) and in danger of descending into madness. His interaction with remains is paradoxically both a way of keeping a grip on the material world, as well as the source of his anxiety, as, on the one hand, remains shape his narrative of self and give meaning to his existence, while on the other, they dramatically undercut his sense of self-presence in the absence that they foreground. The self that develops from reading remains is increasingly experienced by the speaker as a self devoid of substance—as a self that is as bereft of meaning and as lifeless as the remains from which it is constructed. Thus, the speaker cannot avoid madness, as while the escape from the object world (the desire to bury himself in himself) is a symptom of his dissociation from the phenomenal world, the self constructed in his reading of remains is one that is sick and riddled with existential angst.

There are moments in *Maud*, however, when the "dreadful hollow" of the speaker's existence is dispelled; when, for example, remains appear charged with meaning and able to offer full presence. These are the moments when the speaker encounters readable remains that can be reconstructed into self-affirming narratives, as in Part I, when Maud's "happy yes" (I: 579) results in the speaker's heightened emotional excitement:

> From the meadow your walks have left so sweet
> That whenever a March-wind sighs
> He sets the jewel-print of your feet
> In violets blue as your eyes,
> To the woody hollows in which we meet
> And the valleys of Paradise. (I: 888–93)

Love leaves its trace in the form of the "jewel-print[s]" of Maud's feet that are "set" like precious fossil imprints. Like the fossils discussed in Chaps. 1 and 2, the "jewel-print" carries the potential for reconstruction. Mantell discusses what he believed at the time to be the ancient impressions of human feet found in sandstone. The "prints", he writes, present the "perfect impress of the feet and toes, exhibiting the form of the muscles, and the flexures of the skin". A narrative can be read from such remains; stature can be deduced, the individual's "upright" and "easy" stance can be determined.[46] Similarly, the jewel-prints of Maud's feet leaves signs that also yield a narrative; signs that can be followed and that lead to the "woody hollows" and "valleys of Paradise" where the "dreadful hollow" is filled by the living presence of the lovers. The "jewel-print" is ostensibly able to bring the signified into full presence in the reconstruction of the physical woman: the shape of her foot, the violet colour of her eyes. At the same time, however, the "jewel-print" unavoidably points the way to the "hollows" and "valleys" that gesture towards the inescapable absence that the fossil imprint inevitably indicates. Thus, the speaker's faith in the power of remains to recover what is missing is subtly undermined by a more fundamental sense of the vacuity that remains signify. The "jewel-print" is itself another "hollow" like the "dreadful hollow"; it is a space and nothing fills it, the observer merely projects meaning over its void. What the "jewel-print" reconstructs is the speaker's idealised narrative of Maud, while it simultaneously indicates her utter and irrevocable absence.

In another heightened moment, the speaker imagines his own remains resurrected by the power of love, when, in the rapturous final lyric of Part I, Maud is imbued with a sensual capacity to re-animate the speaker's dust:

> She is coming, my own, my sweet,
> Were it ever so airy a tread,
> My heart would hear her and beat,
> Were it earth in an earthy bed;
> My dust would hear her and beat,
> Had I lain for a century dead;
> Would start and tremble under her feet,
> And blossom in purple and red. (I: 916–23)

---

[46] Mantell, *Wonders* (1838), 1: 76.

Here, the speaker's remains are materials for reconstruction as Maud's presence, in an unconscious acknowledgement of the vital role of the "other" in the self's self-awareness, is invested with the power to seemingly bring the speaker into full presence. In the manner of the reconstruction of fossil remains carried out by comparative anatomists, the speaker's "earth" is fleshed out, his "dust" remade, and the heart that was dead (the "poor heart of stone") even while the speaker lived, is dynamically and erotically aroused—flushed through with the purple and red that are the true colours of the heart's veins and arteries.[47]

Such ecstasy, of course, is short lived, and immediately following this episode is the crisis and decent into madness of Part II. Remains now appear as indicators of absence that can be filled only by the observer's gaze. The inability of remains to recover what is missing or dead and the dead world that they signify drives the speaker's impulse to bury himself within himself. Relinquishing the material world for the inner self, however, leaves the speaker disconnected from the temporal and spatial nexus of narrative (the self, in other words, constructed through interaction with the object world) and in consequence, his narrative and his sense of self break down. Just as the object needs an observer without which, as Zimmerman says, it "slips away into oblivion", without remains the speaker loses his narrative of self; he is "blotted out of the book of life" and descends into the "dreadful hollow" of the dissociated self.[48] Here, the temporal and spatial co-ordinates of narrative are dangerously close to disappearing altogether, both for the speaker who experiences various degrees of madness, and for the poem itself, which is so often characterised by critics in terms of its fragmentary form and disjointed plot.

This scenario is played out in *Maud's* shell lyric of Part II (49–77). The speaker's interaction with the shell is perhaps the poem's most overt reconstruction in the mode of comparative anatomy, and, like so much of *Maud*, as Turner notes, it was written in response to Tennyson's

---

[47] That Tennyson had blood in mind when he wrote line 923 is corroborated by James Henry Mangles: "I said I thought he must have meant the heaths, when he wrote 'Blossom in purple and red.' He smiled & said he supposed he meant 'Blood.'" See Mangles, *Tennyson at Aldworth: The Diary of James Henry Mangles*, Earl A. Knies ed. (1984), 40. Also quoted in Shatto 199.

[48] Zimmerman, *Excavating Victorians* (2008), 134. Owen, *Lectures* (1846), II, 3.

reading of Lyell's *Principles*.[49] Remains offer the speaker a narrative hook onto the phenomenal world. Drawn to the shell after the crisis of the ill-fated duel, the speaker contemplates his relationship to the shell.

> Strange, that the mind, when fraught
> With a passion so intense
> One would think that it well
> Might drown all life in the eye, —
> That it should, by being so overwrought
> Suddenly strike on a sharper sense
> For a shell, or a flower, little things
> Which else would have been past by! (II: 106–13)

It is precisely at the point when the intensity of his mental anxiety threatens to "drown" out the phenomenal world that the speaker recognises how the shell, along with other objects ("little things" in comparison to the deep world of the inner self) offer him the possibility of narrative recovery. The shell is what remains of the "living will" and, as such, it provides negative evidence of a past presence that, in the fashion of comparative anatomy, the speaker feels compelled to reconstruct.[50] The compulsion comes from his tacit understanding that it is his narrative of the shell that not only (ostensibly) recovers the "little living will" but that also gives him shape and meaning. The contemplation of the shell results in a tentative narrative. The speaker wonders, did the "little living will/That made it stir on the shore" "stand" at "a diamond door"

---

[49] Turner quotes Lyell, "It sometimes appears extraordinary when we observe the violence of our coast ... that many tender and fragile shells should inhabit the sea in the immediate vicinity of this turmoil" (*PG*, II, 281). See *Tennyson* (1976), 134.

[50] Anderson and Taylor make the connection between the shell lyric and visits Tennyson made to the home of the naturalist and geologist, Charles W. Peach. Peach showed Tennyson "small marine animals" and Anderson and Taylor note that Tennyson later spoke to friends about "lower organisms feeling less pain than higher" and that these visits "foreshadow Tennyson's mid-1850s passage in *Maud*" on the shell and its little living will. See, "Tennyson and the Geologists: Part I" (2016), 350. This connection is good evidence of Tennyson's practical engagement with the natural sciences and suggests the importance of visual as well as textual influences. It also implies that Tennyson was mulling over knotty ideas around the significance of pain and emotion for animals, and the commonality or gradation of emotional susceptibility between animals and humans.

in a "rainbow frill"; did it push a "golden foot or a fairy horn" through a "dim water-world" (II: 62–68). Hugh Miller's provocative language seems to surface here. Miller laments the geologist's inability to reconstruct the colour of the world of fossil fish—the "'waved coats, dropt with gold,' the rainbow dyes of beauty of the watery tribes"—because, he writes, "color is a mighty matter" and is "connected often with more than external character". It is a "curious and interesting fact", Miller goes on to suggest, "that the hues of splendor in which they are bedecked are, in some instances, as intimately associated with their instincts—with their feelings, if I may speak so—as the blush which suffuses the human countenance is associated with the sense of shame, or its tint of ashy paleness of sallow with emotions of rage, or feelings of a panic terror" (*ORS*, 251). The speaker takes the task of reconstruction further than the comparative anatomist, animating the empty shell and filling it with the rich spectrum of emotion projected from his numb and alienated self, and, with the recovery of the "living will", comes the recovery of the speaker's sense of himself as a player in the object world, which enables him to "strike on a sharper sense". Like the "Echo" (I: 4) in the "dreadful hollow", his narrative of the empty shell rebounds back to him to fill the "hollow" of his own self. However, as his own voice is reflected back to him from dead remains, the narrative of self thus constructed ultimately only compounds his sense of emptiness and meaninglessness. The speaker's narrative cannot recover the "living will", and thus, like the voice of Echo, such narratives speak only of "Death" (I: 4). The paradox is that while the shell allows him to "strike on a sharper sense" of self and to momentarily surface from the objectless "dreadful hollow", it is an object that speaks to him of his own future as remains, and of the dead world of meaningless objects to which he himself and all life must ultimately submit.

The validity of the speaker's narrative is brought into question when, in the same moment of reflective clarity that allowed him to "strike on a sharper sense" of self, he recalls Maud's dying brother:

> And now I remember, I,
> When he lay dying there,
> I noticed one of his many rings
> (For he had many, poor worm) and thought
> It is his mother's hair. (II: 114–18)

The hair fashioned into a memorial ring is another example of organic remains that keep the speaker anchored in the object world by offering him materials for narrative construction. It recalls the lock of hair in *The Princess* that reshaped Ida's narrative from one of "great deeds"—the footprint hardened into stone—to one of reproduction, mortality and domesticity. Here, however, it is an object that has the potential to disrupt the general rhetorical thrust of the poem as it problematises the speaker's rendering of Maud's brother as a "dandy-despot" (I: 231), a "Sultan" (I: 790) and a "oil'd and curl'd Assyrian bull" (I: 233). Hair jewellery, as Deborah Lutz points out, "had its own narrative", and the sentimental ring feeds into Maud's own account of her brother as "rough but kind", and as her steadfast "nurse" (I: 759).[51] It speaks of the possibility of other narratives, of, for example, the affecting narrative of a son's love for his mother. The ring is also only one of a number ("For he had many"), suggesting that any number of potential narratives threatens to disrupt the speaker's shaping of a history in keeping with his perception of events. Like all memorial relics, a ring of hair may have had "value only to a handful of people, or even to just one, and if that one died, then the relic became ... of worth to no one".[52] With the death of the brother, both mother and brother slip further into the past. The ring is no longer a trace to be interpreted by the brother and it cannot be read as the brother once read it. It is noticed by the speaker, yet the meaning invested in it by Maud's brother is no longer available. The ring instead encircles a vacant space, another "dreadful hollow" upon which an observer may project any number of meanings, while its original significance is wholly lost. The speaker's contempt for the brother's many rings betrays his unreliability as a reader of remains and serves to demonstrate the ambiguity of the object and the subjective and arbitrary nature of interpretation.

Remains present the speaker with the unthinkable: the absence of self. They serve a dual role in bringing the speaker (his narrative of himself) into being, while they also poignantly demonstrate the condition in which consciousness faces both the impossibility and the inevitability of non-existence. In the madhouse canto of Part II, the speaker projects his consciousness onto his own future remains envisaged as subsumed into

[51] Deborah Lutz, "The Dead Still Among Us" (2011), 128.

[52] Ibid., 129.

the geological system and embedded like a fossil in strata beneath the city streets.

> Dead, long dead,
> Long dead!
> And my heart is a handful of dust,
> And the wheels go over my head,
> And my bones are shaken with pain,
> For into a shallow grave they are thrust,
> Only a yard beneath the street,
> And the hoofs of the horses beat, beat,
> The hoofs of the horses beat,
> Beat into my scalp and my brain,
> With never an end to the stream of passing feet,
> Driving, hurrying, marrying, burying. (II: 239–50)

In this lyric, the speaker's predilection for the "selfish grave" (I: 559) is realised in his burial in the "suicide's grave at the cross-roads of the street" and, like his father, who made "false haste to the grave" (I: 58), the speaker's "bones" do not undergo any common process of decomposition, rather they are beaten and flattened into the ground in actions that once again draw attention to processes of fossilisation.[53] The "hoofs of the horses [that] beat,/Beat" upon the speaker's "scalp and brain", the incessant action of the "wheels" and the never-ending "stream of passing feet", all perform the same flattening, crushing, dinting actions that, in Lyellian manner, appeared to divest the father's remains of its character and class. The living world, as in Carlyle's description, "press[es] upon the lower, squeezing them ever thinner", removing the dead from the reach of history.[54] Thus, while the speaker "cr[ies] to the steps above" (II: 339) his head, he cannot be heard, as his dishonourable remains, like his father's, leave no readable trace.

The speaker's fate recalls Lyell's and Mantell's extraordinary extrapolation of geological processes that attempt to show in what ways the human remains of the present world may come to be integrated into the cities of the future. Incorporated into the world of inorganic matter, the

---

[53]William B. Thesing, "Tennyson and the City: Historical Tremors and Hysterical Tremblings" (1977), 17.

[54]Carlyle, *Historical* (1898), 64.

speaker enters the "ghastly pit" of geological time and, fittingly, from this position human time is vastly accelerated. The business of human existence appears as a frenzied rush of "driving, hurrying, marrying, burying", a flickering and relentless succession of births and deaths.[55] The speaker's bones, like the "skeletons of men" that Lyell suggests are "stamped in solid rock" (1: 360), are beaten into the city streets to form yet new strata upon which the fleeting lives of future generations play out.

Already "Gorgonised", his heart "half-turn'd to stone", the speaker's inhumation in the strata completes the process of fossilisation, all but for the fact that, in this terrible state of being, he remains conscious. The "rough grave" is not the longed for "still cavern deep" (II: 236) where the speaker envisaged his troubled consciousness stilled by the geological processes that would fix it into stony breccias. As he laments, "I thought the dead had peace, but it is not so" (II: 253). Rather, as Lyell's text suggestively states, in the process of fossilisation, remains are "imprisoned in solid strata [where] they may remain throughout whole geological epochs before they again become subservient to the purposes of life" (*PG*, II, 189). The real life of matter, it would seem, begins with death, and, in keeping with *Maud*'s internal logic, while the speaker's consciousness lives on, the living men that occupy the city appear to be already dead; "Ever about me the dead men go /And then to hear a dead man chatter /Is enough to drive one mad" (II: 256–58). In a world devoid of spiritual meaning, where the "churches have kill'd their Christ" (II: 267), where coal and gold are valued above the living individual, where men are fed with "chalk" and "alum", truly "only the ledger lives" (I: 35). The appearance of life—the "chatter" and "babble" of men—belies an absence of soul suggesting the living body is already bereft of meaningful life. Thus, *Maud* envisages a geologically oriented nightmare in which the morally and spiritually dead confront an afterlife that imprisons them within their own material remains. With no soul to transcend the body, consciousness becomes enmeshed within processes of fossilisation in a horrifying parody in which divine eternal life becomes instead a material damnation of perpetual geological process.

---

[55] This image owes much to Lyell, who, in order to postulate deep time, speeds up human time. See, for example, *PG*, I, 78–79.

THE SPEAKER'S READABLE REMAINS

Sifting through *Maud*'s remains, the speaker finds meaning to be merely the reflection of the observer's subjectivity over blank materiality. Part III, however, attempts to fill the absence encoded in the "dreadful hollow" and the "ghastly pit" by restoring faith in the existence of extrinsic meaning—by investing, in other words, in a narrative that ostensibly transcends the self: a shared narrative that can indelibly imprint itself on the speaker's remains and will be legible for all time. Such a narrative is found in the rhetoric of nationalism and war and the rallying cry that sees the "heart of the people beat with one desire" (III: 49). War answers the speaker's specific concerns about his own remains and what may or may not be read from them. It promises to confer upon his remains a narrative of honour and heroism, thus avoiding the suicide's grave and the dishonourable fate of the father's remains.

Aptly, it is Lyell's geological text that offers a solution to the problems it initiated. Lyell investigates "in what manner the mortal remains of man and the works of his hands may be permanently preserved" (*PG*, II, 253) and highlights war as offering a unique pathway into the fossil record. Perishing at sea, Lyell suggests, offers particularly favourable conditions for the preservation of the human form, and specifically for those who go down in the wreck of a man-of-war. In these cases, "cannon, shot, and other warlike stores, may press down with their weight the timbers of the vessel when they decay, and beneath these and the metallic substances the bones of man may be preserved" (*PG*, II, 256). Crucially, it is the paraphernalia of war itself that guarantee the immortality of human remains, as the bones of fighting men are flattened, compacted and preserved under the pressure of sunken weaponry.

Appropriately, those who perish are "crush'd" (III: 44), like the speaker's father, in language foregrounding actions conducive to fossilisation. And yet, unlike the ignominious fate of the father's remains, remains found in these circumstances speak distinctly and specifically of what is in *Maud*'s rhetoric a self-sacrifice that bestows the highest honour. Lyell notes that "During our last great struggle with France, thirty-two of our own ships of the line went to the bottom ... In every one of these ships were batteries of cannon constructed of iron or brass, whereof a great number had the dates and places of their manufacture upon them in letters cast in metal" (*PG*, II, 256). Such artefacts not only leave lasting and unequivocal narratives literally written into the materials of war, in Lyell's envisaging of men thus entombed, those narratives are imprinted and incorporated into their remains. Thus, for the speaker, war fills the empty space of his imagined future remains with a narrative that writes itself into the heroic body, preserving it for all time. Lyell "anticipate[s] with confidence" that those fortuitous remains crushed beneath the artefacts of war "will continue to exist when a great part of the present mountains, continents, and seas have disappeared" (*PG*, II, 271). War bestows its seemingly meaningful narrative, not only on the speaker's remains but on the remains of the age, offering a type of immortality that for the speaker tells of the "Honour that cannot die" (I: 177). This final illusion ends *Maud*, as the speaker stands on the "giant deck" of a warship bound for the "Black and the Baltic deep", for "battle, and seas of death" (III: 34, 51, 37), and, it would seem, for a very specific form of immortality in the fossilisation and preservation of the speaker's own heroic and eminently readable remains.

CHAPTER 6

Maud and the Unmeaning of Names: Geology, Language Theory and Dialogism

Daily it is forced home on the mind of the geologist, that nothing, not even the wind that blows, is so unstable as the level of the crust of this earth. (Charles Darwin)[1]

Maud's interrogation of 'remains' suggests Tennyson's nuanced engagement with the images and ideas that geology provoked. However, as in *In Memoriam*, *Maud*'s geology is also uniformitarian. Where *In Memoriam* both experiments with, and strenuously resists, the deep implications of the uniformitarian vision, *Maud* takes it on fully at the risk of complete collapse. *Maud* represents the climax of Tennyson's uniformitarian poetics and it demonstrates, perhaps better than any other text of the period, poetic or otherwise, the paradigmatic shift that uniformitarianism triggered. *Maud* has had a chequered critical reception, with many contemporary critics regarding the poem as odd, disturbing and even offensive. Reactions to the poem point to how *Maud* had a way of drawing its readers into a dialogue that, it seems, they neither anticipate nor very often welcome. Modern critics have also been quick to recognised *Maud*'s variance and responses to the poem have been rich and diverse. This chapter argues that the poem registers a dialogic perception on a number of levels. Crucially, however, not only does the poem's dramatic action rest on a vivid geological tropology (part of which has already been

---

[1] Charles Darwin, *The Voyage of the Beagle* (1997), 305.

M. Geric, *Tennyson and Geology*, Palgrave Studies in Literature, Science and Medicine, DOI 10.1007/978-3-319-66110-0_6

discussed in Chap. 5), its linguistic structure is also geologically oriented. What this chapter hopes to show is that *Maud*'s dialogism is the product of a subtle extrapolation of Lyell's uniformitarianism across the fields of language theory—an extrapolation that profoundly disrupts conventional mid-nineteenth century perceptions of language as Adamic and teleological.

To appreciate fully Tennyson's understanding of the cultural impact of Lyell's text, his knowledge of Whewell's theologically orientated discourses on geology, as discussed in Chap. 3, need also be taken into account. Whewell's texts served as sites of intersection where debates in geology and language theory were tested out in terms of their adherence to Whewell's sophisticated, albeit idiosyncratic, conservative orthodoxy.[2] Among the other discourses that echo in *Maud*, Robert Chambers's provocative *Vestiges of the Natural History of Creation* (1844) and Richard Chenevix Trench's *On the Study of Words* (1851) are perhaps the most noteworthy. Isobel Armstrong alerts readers to the profound significance of both Lyell's and Trench's thinking for the dynamics of *In Memoriam*; by extending Armstrong's discussions to *Maud*, the poem can be read in terms of contemporary debates on the origin of language—debates that were, in the case of Whewell and Chambers, fittingly embedded within geological discourses.[3]

As argued in the first chapter, literature and science of the first half of the nineteenth century did not labour under the present terms of disciplinary division and therefore cannot be understood from the perception of the 'two cultures' debate. Moreover, implicit in this dialogic reading of Tennyson's writing is the rejection of a unidirectional model of textual influence. Thus, *Maud* is not read as merely affected by geological theories and discourses on language, but as a text that performs those debates and discourses and brings them into being in its dialogic experimentation. Dialogic discourse, as Michael Macovski asserts in his valuable volume of collected essays on Bakhtinian interpretation, "includes not only the interchange of voices within texts, but interchange *between* texts as well—across discourses separated 'in time and space'".[4] Just as

---

[2] For a thorough account of Tennyson's understanding of language theory, see Donald S. Hair, *Tennyson's Language* (1991).

[3] Isobel Armstrong, *Victorian Poetry: Poetry, Poetics and Politics* (1993), 256.

[4] Michael Macovski, "Introduction" to *Dialogue and Critical Discourse: Language, Culture, Critical Theory*, Michael Macovski ed. (1997), 8.

Maud draws readers past and present into an often argumentative dialogue, it also re-energises with each reading the convergence of intertex-
tual enquiries, debates and rejoinders. The aesthetic form is the conduit
of meaning; it permits play between utterance and reply and allows for
the continual production and reproduction of meaning. Therefore, to
speak of *Maud* as influenced by debates in geology or linguistics is to
miss the crucial role that the poem plays in conferring meaning on these
texts by giving them cultural context, as Chap. 1 suggested. Thus, it is
not only that *Maud* offers the possibility of recovering a largely overlooked dialogue between emerging sciences (geologic and linguistic);
more profoundly, *Maud* is that dialogue. The poem enacts in cultural
praxis the linguistic crisis that Lyellian geology initiated. It does this by
experimenting in a poetics that is both anti-monologic and radically anti-
teleological. E. Warwick Slinn's compelling analysis of Victorian poetry
discusses the poetic expression of the "movement from Hegelian difference—the prospect of the unity of consciousness that is constituted
through division—to Derridean *différance*—the unity of closure that
is continually deferred".[5] This movement is strikingly played out in
Tennyson's *Maud*, as Bakhtinian theory helps to show. *Maud* goes further than *In Memoriam* in the realisation of a uniformitarian consciousness; it takes the trajectory of Lyell's geologic method to its ultimate
conclusion for the monologic mindset—to, in other words, the precipice
of madness. Such a reading of the poem is indebted to, and I believe,
entirely compatible with Slinn's Hegelian reading of *Maud*. Slinn per-
ceptively shows how "within the Tennysonian present, completion and
fulfilment are not possible".[6] In *Maud*'s experimentation with dialogism,
this impossibility is realised through a speaker whose consciousness is
depicted as processive and unfinalisable. Further to this, however, *Maud*
offers a textual manifestation of the confrontation between the centripetal pull of teleological epistemologies and the centrifugal push of heteroglossia. In its geological dialogics, it records an epiphanic moment
of crisis that pre-empts post-structuralist and postmodernist encounters
with the unstable subject in language.

---

[5] E. Warwick Slinn, *The Discourse of Self in Victorian Poetry* (1991), 8.
[6] Ibid., 64.

DIALOGISM

One dialogic approach to Tennyson's poetry has come from Saverio Tomaiuolo, who sees *In Memoriam* and *Maud* as discourses "in which various ideological positions are clashed and fused together to compose, in Mikhail Bakhtin's words, a 'heterology'". However, Tomaiuolo suggests that *In Memoriam* might be "figuratively imagined as a centrifugal poem based on a semantic expansion from an all-pervading narrative voice", while *Maud* "is a centripetal poem where all the phenomena surrounding the speaker are reduced to minute detailing, subjected to the restricted focus of a deranged mind".[7] I would argue, however, for a somewhat different view, as in Bakhtinian terms, centripetal forces are centralising forces; they are the socio-ideological forces that attempt to submerge heteroglot plurality under the hegemony of official discourses and that work to produce a sense of monologic unity out of the heteroglossia of social discourse. If *Maud* is a "centripetal poem", it must necessarily be seen as tending towards social cohesion—as based within, and gravitating increasingly towards a centralised ideological position. This surely more accurately describes *In Memoriam*, where the speaker's inner dialogue articulates his desire to assimilate death into theologically framed ideologies, those official teleological discourses that characterised mid-Victorian sensibilities. In *In Memoriam*, as Chap. 4 tried to show, the speaker battles the centrifugal forces that work to decentre him, finally reasserting his linearity and re-affirming an overall sense of monologic unity. Thus, it is *In Memoriam* rather than *Maud* that appears to be centripetal.

In contrast, *Maud*'s stylistic fragmentation, along with its theme of madness, indicates a breakdown of monologic unity. *Maud*'s speaker exists outside the privileged ideological centre represented by the manorial hall, and his rejection of the demands of wider social structures, the "morbid hate and horror... / Of a world in which I have hardly mixt" (I, 264–5), suggest acute alienation. The speaker's desire to bury himself in himself is not "centripetal"; as in *Maud* subjectivity does not signal a monologic fortification of inner coherence—it signals difference, as the more idiosyncratic the speaker's voice, the more marginalised and decentred he realises himself to be. In one sense then, *Maud*'s speaker is exposed to centrifugal force: the force of heteroglossia, which breaks

---

[7] Saverio Tomaiuolo, "Tennyson and the Crisis of Narrative Voice in *Maud*" (2002), 29.

down ideological thought into multiple positions. The speaker's madness is the result of his loss of monologic certainty and his realisation that a social nexus confronts and addresses him and demands a dialogue—a dialogue from which he recoils. More accurately, perhaps, *Maud* is neither centripetal nor centrifugal; rather, the poem is an experiment in the texture of dialogism, manifesting the pull and push of centripetal and centrifugal forces through a speaker who has passed from a monologic perception of the world (the world view of *The Princess*) to one that forces him to recognise a multiplicity of ideological positions. *Maud*'s representation of heteroglossia, figured in the "idiot gabble" (II, 279) of the 'madhouse canto', allegorises a post-teleological realisation that language is not monoglossic but multi-accented, that individual consciousness is not monolithic but dialogic, and that the authored/authoritative text is not absolute but provisional, mutable, movable in terms of time and space and therefore infinitely interpretive. Such a dialogic reading is compatible with Herbert Tucker's seminal and perceptive analysis of *Maud*'s negotiation of the divisions between public and private spheres. Tucker suggests that what "assails" *Maud*'s speaker in the "babble" of the poem's second section is "rather an excess than a dearth of meaning" as the speaker confronts not only a "breakdown between public and private discourses" but also, in his apprehension of language as social sign, the dissolution of monologic sense and teleological certainty.[8]

In line with the reversal of Tomaiuolo's argument, *Maud*'s "minute detailing", the "hyper-detailed description of objects, physical attributes and events" that Tomaiuolo sees as suggestive of the "restricted focus of a deranged mind", can be read alternatively as the unrestricted vision of a consciousness that sees everywhere infinite multiplicity.[9] *Maud*'s visual detail is profuse, as exemplified in the speaker's observation of the "eyelash dead on the cheek" (I, 90), the "ring" fashioned from a lock of hair (II, 116–8), the lavish description of the tiny "shell" (II, 49–77) and the microcosmic world of the "little wood" (I, 125). Such attention to minutiae eerily evokes the astonishing detail of Richard Dadd's painting *The Fairy Feller's Master-Stroke* (1855–64). Like *Maud*'s mad speaker, Dadd, insane and incarcerated for patricide, finds significance in the smallest of elements, producing a work of stupefying visual depth that suggests both microscopic infinity and a myriad of interpretive

---

[8] Herbert F. Tucker, *Tennyson and the Doom of Romanticism* (1988), 427.

[9] Tomaiuolo (2002), 29.

possibilities: a madness that sees meaning everywhere. *Maud*'s depiction of detail indicates a visual overload rather than a reduced vision. In Bakhtinian terms, it gestures towards the unfinishability of consciousness and of the difficulty of shutting out the detail and ambiguity that impedes a monologic sense of the world.

This chapter returns to a close reading of *Maud*'s formal engagement with dialogism later. However, the connection between discourses in geology and language theory, touched on in Chap. 1, requires some historical contextualisation as it is these connections that *Maud* registers and that work to produce the poem's experimental dialogism.

GEOLOGY AND LANGUAGE THEORY

The radical potential of Lyell's uniformitarianism, as *In Memoriam* suggests, is found in its prioritising of present change and its anti-teleological disregard for first and final causes. The uniformitarian methodology, as already explained, postulated that the past was best understood via observation of the present. For William Whewell, however, the emphasis on uniformitarianism—useful as it was as a methodology—suggested a dangerously radical move away from Adamic and teleological epistemologies. Present processes, under uniformitarianism, are seen as the major agents of change, making everything in the present appear in flux. Stable categories—those hard-won inviolable taxonomic divisions that created meaning and that ensured the stability of the hierarchical order for the Victorians—were dissolved by the uniformitarian focus on the present, as Darwin saw. This radical potential was not lost on Whewell or Tennyson, who both recognised the wider cultural and epistemological implications of the triumph of uniformitarian thinking over catastrophist thinking. However, for Tennyson, it is not until he comes to write *Maud* that he is able to take the implications of Lyell's uniformitarianism to its limits. The uniformitarian focus on continual change in the present is manifested in *Maud*'s geological tropology, which figures fluidity rather than fixity. Images of petrifaction and the transformation of organic material into an inorganic state are ubiquitous in *Maud*. The poem's characters exist in a uniformitarian nightmare in which the slow processes of geological change occur in human rather than deep time (the subject of Chap. 4). Uniformitarian process provides the backdrop against which the personal 'catastrophic' events that move *Maud*'s plot—the father's suicide, the climactic duel—are played out. Thus, just as Lyell sets geological catastrophe against the more subtle and profound forces of uniformitarian geological change, *Maud* figures the personal catastrophes that befall the poem's characters against a set of impersonal, hardly perceptible agents of change that work insidiously and uniformly across *Maud*'s landscape. Images of petrifaction articulate a number of anxieties; on one level, they express concerns over the reification of human values and therefore constitute part of the poem's critique of *laissez-faire* industrial capitalism as devoid of social policy, bereft of a fleshy heart. However, they also express a fear of taxonomic instability in both geological and in linguistic terms. In this, Tennyson is responding not only to Whewell's reaction to Lyell but also to Adamic accounts of the origin of language and arguments concerning the dependence of knowledge on the fixity of types—polemics exemplified not only in the work of Whewell but also in that of Trench and Chambers.

The conventional post-Enlightenment consensus was to see language as divinely originated. Adamic philologists postulated that the further back (in etymological terms) words could be traced the nearer they approach their divine and perfect root. Thus, as linguistic historian Hans Aarsleff notes, through etymology Adamicists "sought to recapture lost perfection and divine authority".[10] Chambers challenged Adamic doctrine in *Vestiges* in 1844 by assuming a basically uniformitarian premise and arguing that observable changes to language taking place in the present could explain the development of language in the past. Thus, he postulated a material origin for language, asserting that it had developed in parallel with the progress of humankind from "barbarism" to "civilisation". Language is "so remarkable", Chambers writes, "that there is a great inclination to surmise a miraculous origin for it ... Here, as in so many other cases, a little observation of nature might have saved much vain discussion."[11] For Chambers, language development "was as natural and uniform as geological formations, its understanding independent of indications of design in the Creator". Thus, *Vestiges* "pushed language

[10] Hans Aarsleff, *From Locke to Saussure: Essays on the Study of Language and Intellectual History* (1982), 31.

[11] Robert Chambers, *Vestiges of the Natural History of Creation* (1844) Facsimile reprint (1994), 310–311.

into the center of the discussion" with an argument "chiefly erected on astronomical and geological evidence".[12]

Whewell replied to Chambers in his *Indications of the Creator* (1845), enquiring; "what glossologist will venture to declare that the efficacy of such causes has been uniform; that the influences which mould a language ... operated formerly with no more efficacy than they exercise now ... ?" Whewell concludes, "we cannot place the origin of language in any point of view in which it comes under the jurisdiction of natural causation at all".[13] Whewell had an enduring interest in language. He was a member of the Cambridge Etymological Society and the Philological Society of London, contributing many key terms to scientific terminology. Significantly, his rebuttal of Chambers and Lyell came in his discussion of geology in a section headed the "Doctrine of Catastrophes and Uniformity".[14] Thus, Chambers and Whewell (both of whom were read by Tennyson) embed a debate on the origin and development of language within geological discourses and specifically those centred on the antagonistic relationship between catastrophism and uniformitarianism (see Chap. 3).[15]

Despite Chambers's assault on Adamic doctrines, they survived most notably in the work of Trench, who was still peddling—what linguistics historian Bernd Naumann tellingly calls—a "catastrophistic" doctrine of language, as late as the 1850s.[16] Trench's texts, as Tony Crowley asserts, were the "major British work on language in the 1850s" and his approach to the English language was to see it as a "support for theological dogmatism".[17] Trench was a Cambridge Apostle at the same time as Tennyson, although they disagreed temperamentally. The prickly relationship between the two men was probably behind Trench's indirect dig at Tennyson's inaccurate (according to Trench) use of 'mellay' and 'medley' in *The Princess*, as suggested in Chap. 1. Trench was the

---

[12] Hans Aarsleff, *The Study of Language in England 1780–1860* (1983), 226, 223.

[13] Whewell, *Indications* (1845), 164, 168.

[14] Aarsleff, *Language* (1983), 224.

[15] Cymbre Quincy Raub (1988), 287–300

[16] Bernd Naumann, "History of the Earth and Origin of Language in the 18th and 19th Century: The Case for Catastrophism," in *Language and Earth*, Bernd Naumann, Franz Plank and Gottfried Hofbauer eds. (1992), 49.

[17] Tony Crowley, *The Politics of Discourse: the Standard Language Question in British Cultural Debates* (1989), 52, 55.

one Apostle "not enthusiastic" about Tennyson's *Poems, Chiefly Lyrical* (1830), and in turn, Tennyson flippantly professed to "have no faith in any one of his opinions" and wrote of Trench (in what was clearly a gibe at Trench's overbearing piety) "the earnestness that burns within him so flashes through all his words and actions, that when one is not in a mood of sympathetic elevation, it is difficult to prevent a sense of one's own inferiority and lack of high, and holy feeling".[18] Such earnestness burns through the pages of Trench's *On the Study of Words* (1851). Language for Trench was a manifestation of an innate, God-given faculty: "God gave man language, just as He gave him reason, and just because He gave him reason, (for what is man's word but his reason coming forth, so that it may behold itself?)".[19] This is not to say that "man started at the first furnished with a full-formed vocabulary of words ... He did not thus begin the world *with names*, but with the *power of naming*". Thus, it is "not God who imposed the first names on the creatures, but Adam ... at the direct suggestion of his creator" (*Words*, 15). This leads Trench to argue that although the "divine idea" is innate, fixed and immutable, there may be multiple linguistic variations through which it can be expressed. New coinages in language were merely a further dissemination of a "divine idea", a dividing and sharing of the original sense of Adamic words, as all divergences "may be brought back to the one central meaning, which knits them all together" (*Words*, 188–189). Language degenerates when ambiguity enters into play.[20] Trench urges the individual to interrogate words and isolate their meanings: "Learn to distinguish between them" he writes, for "the mixture of those things by speech, which by nature are divided, is the mother of all error'" (*Words*, 177).

For Trench, the boundaries that enclose word groups are absolute and any violation of the integrity of the original meaning of a word can only result in the loss of the original God-given idea. An example is Trench's discussion of the word 'plague'; "There are those who will not hear of great pestilences being God's scourges of men's sins; who fain would find out natural causes for them", yet the word plague, Trench

---

[18] Robert Bernard Martin, *Tennyson the Unquiet Heart* (1980), 112. Also see *Letters ALT*, A.T. to William Henry Brookfield, mid-March (1832), 71.

[19] Richard Chenevix Trench, *On the Study of Words* (1851) (1853), 14–15. Hereafter cited parenthetically as *Words*.

[20] See Isobel Armstrong's discussion of *In Memoriam* and the "Trenchian requirement to repress double meaning" *Victorian Poetry* (1993), 257.

insists, means "according to its derivation, 'blow,' or 'stroke'; and was a title given to ... terrible diseases, because the great universal conscience of men, which is never at fault, believed and confessed that these were 'strokes' or 'blows' inflicted by God on a guilty and rebellious world" (*Words*, 37–38). Thus, when disease is attributed through science to natural causes, the result is the loss of the 'divine idea' which in Adamic theory is self-evident in the etymological root of the word. A similar "source of mischief", Trench asserts, is found in the "practice of calling a child born out of wedlock, a 'love-child', instead of a bastard" (*Words*, 49).

Trench's intransigence is part of a reactionary fear of the mutability of language and the potency of speech in the shaping and reshaping of meaning—facets of language that Adamic theory strenuously attempts to suppress. Trench's rhetoric exemplifies how the "search for linguistic unity and identity is one that is founded upon acts of violence and repression: a denial of heteroglossia—discursive and historical—in favour of centralising, static forms".[21] To secure the boundaries of meaning, language must always look back to the authority of the past, words must always refer back to a presumed fixed and original 'truth'. Trench attempts to rein in meaning and to eradicate ambiguity in an effort to safeguard a set of conservative ideologies by securing what is, in Bakhtinian terms, a centripetal hold on language's inherent tendency to change with social usage. If word meanings are seen to be mutable and ever-changing in the present (the uniformitarian view), then the hierarchy of the "divine idea" over the word would be reversed, resulting in (what might be considered in Trench's Adamic system of thought) a perverse parody of divine language.

Trench's fear of the "mixture of those things by speech, which by nature are divided" betrays a general conservative fear of mutability both linguistic and biological. Mutability is as unnatural in language as evolution is in biology, and both threaten teleological epistemologies by dissolving taxonomic boundaries and promoting in their place continuous change in the present. Similarly, Whewell's philosophical horizon could not entertain evolutionary or transmutive epistemologies, whether geological, biological or linguistic, because, as Aarsleff claims, "Whewell like Cuvier felt that the fixity of types was a pre-condition for knowledge".[22]

---

[21] Tony Crowley, *The Politics of Discourse* (1989), 9.

[22] Aarsleff, *Locke to Saussure* (1982), 35.

Whewell writes, "Objects are so distinguishable by resemblances and differences, that they may be named, and known by their names. This principle is involved in the idea of the Name; and without it no progress could have been made." Progress is here, as it is for Trench, the teleological fulfilment of immanent design. Whewell goes on to assert:

> Intelligible Names of kinds are possible. If we suppose this not to be so, language can no longer exist, nor could the business of human life go on. If instead of having certain definite kinds of minerals, gold, iron, copper and the like, of which the external forms and characters are constantly connected with the same properties and qualities, there were no connection between the appearance and the properties of the object;—if what seemed externally iron might turn out to resemble lead in its hardness; and what seemed to be like gold during many trials, might at the next trial be found to be like copper; not only all the uses of these materials fail, but they would not be distinguishable kinds of things, and the names would be unmeaning.[23]

If entities become indistinguishable, if fixity is replaced by gradation, language can no longer mean, names fail and an epistemological crisis follows. Of course, the examples Whewell gives are solid enough, gold, iron and copper can be relied upon to act consistently in ways that confirm their distinguishability. Nevertheless, Whewell's discourse conjures up, by default, a surreal world of melting and merging entities, of dreadful impermanence and ambiguity. This was a world that Tennyson had already encountered in his reading of Lyell, and which he describes in *In Memoriam*'s vision of the "hills" as "shadows" (CXXIII, 5–8) and in *Maud*'s depiction of processes of fossilisation occurring in the present. Whewell's discourse raises the questions: What if such forms are less than stable? What if these seemingly solid materials proved to be mutable?

Maud manifests Whewell's nightmare. Not only are individuals transmogrified in the text's speeded up vision of petrifaction, all materials, even gold, exist as interchangeable and subtly mutable. Coal, which is at the core of the poem, fuels the rise of the "new-made lord" (I, 332) initiating the subsequent economic inequalities that energise the poem's dramatic action, as well as figuring significantly in its invective against industrial economics. On a geological level, coal signifies the long lineage

---

[23]William Whewell, "On the Fundamental Antithesis of Philosophy" (1849), 8: 177.

of the organic past. In this tropology, the "gutted mine" (I, 338) fore-grounds both the disruption of the natural order of geological strata, and the violation of an "age-old" feudal order, as what has lain in the earth for aeons as the permanent asset of the landed classes is appropri-ated by a new and insatiable class of industrialists and rapidly consumed within the factory furnace. Quick profits are made from the excavation of coal, which conversely has taken millions of years to form. The land is revalued as a capital commodity and hollowed out from under the very feet of the landed classes, literally under*mining* hereditary rights of own-ership and justifying the speaker's fear that the "solid ground" might "fail" beneath his feet (I, 398–99). Ironically, the presence of coal in the British landscape suggests a teleological link between the ancient creation of strata and the present economic wealth of the country. Coal, it seems, is placed beneath the feet of the landed classes to one day make Britain rich. However, coal is a finite resource, and the "gutted mine" also stands as a representation of how the "true" wealth of the aristocracy is improperly exchanged for hard cash. The destabilised landscape reflects a destabilised social hierarchy in which the speaker's fall from social stand-ing (his banishment from the "high Hall") mirrors the rise of the "new-made lord" (I, 332). Thus, in the rhetoric of the text, nobility acquired through generations of breeding is exchanged for "a padded shape" (I, 583). Another more ominous exchange is also implied by the "gutted mine", as the life of the miner who labours in "a poisoned gloom" (I, 337) is exchanged for the precious minerals therein. In this disturbing trade, living flesh is exchanged for dead remains, the carbon deposits of the deep past.

To complete this chain of exchange, coal is subsequently exchanged for gold. The "coal all turned into gold" (I, 340) alludes to the absurd exchange of one lifeless geological material for another. Encoded in this exchange is a critique of what the poem sees as an imprudent industrial rapacity that dissolves time-honoured distinctions as coal, gold and human flesh lose their distinguishability and merge under a single economic equivalence. Thus, everything in *Maud* appears as on a sliding scale, and the pattern for this mutability is established early in the poem. The "fly-ing gold of the ruined woodlands" (I, 12) vividly expresses a sense of the transformative quality of matter and captures within a remarkably small textual space the incomprehensible vastness of geological time. The sea-sons that mark the passage of human time are here aeons, as the onset of winter prefigures the geological transformation of the woodlands into the coal beds of a future landscape, which in turn (and in line with the economic structure of the poem) is "all turned into gold". Time truly "flies", as the wind-blown autumn leaves of an already "ruined" woodland turn to gold before they reach the ground. In this vision of change the woodland is proved to be, as the speaker later claims, genuinely "dearer than all" (I, 887). It is a distinctly Lyellian and uniformitarian image in which deep time brings into focus the gradual processes that show matter to be continuously mutable, and mutability to be the only law.

In *Maud*, the "properties" and "qualities" of entities appear no longer to be inscribed by "form" and "character", indicating the dissolution of "distinguishable kinds". And under such conditions, Whewell tells us, "names [… are] unmeaning", the world unknowable, and "language can no longer exist". Trench, as already suggested, also warns his readers not to mix "those things by speech, which by nature are divided", but if there is no distinguishability in nature, language (at least in its Adamic conception) ceases to mean. Rather than the "idea" being immanent in the entity, awaiting designation through the God-given power of naming vested in man, here there is no prior language, no teleological promise embedded in words; the hierarchy of "idea" and "word" is destroyed, and language becomes dynamic and perpetually mutable in the present. This is for Trench the Godless world in which "love-child" replaces "bastard", and in the process, fixed ideologies become flexible negotiations. *Maud*'s speaker, banished from the ideological centre—the site of the production of meaning represented in the "old manorial hall" (II, 220)—and convinced he alone sees the creeping petrifaction that grips those around him and threatens to incorporate them into the geological system of dead nature, experiences language as devoid of teleological meaning. Buried beneath the city streets, he confronts language at street level, at the crossroads that is at once both the suicide's grave that offers no peace and the site of border crossings and encounters—the mediating chronotope that is the hub of dialogic exchange. Here the "Clamour and rumble, and ringing and clatter", the "idiot gabble" and the "babble merely for babble" (II, 251, 279) is a heteroglossic nightmare of disorderly and unfixed signification.

Maud answers Whewell's questions: what if iron turned out to resemble lead in its hardness? What if gold was found to be like copper? By figuring a uniformitarian world where mutability and infinite gradation is the law; a world in which "distinguishable kinds" do fail, and with them, the taxonomic grounds of all knowledge. In consequence, just as *Maud*'s "gutted mine" is emptied of its buried treasure, language is emptied of its teleological meaning, words become empty signs represented by the "dreadful hollow" (I, 1) and speech becomes "unmeaning"—"Nothing but idiot gabble!" (II, 279). Thus, if meaning in language depends on fixity, Whewell's ostensibly solid and distinguishable geological materials make poor examples and fittingly, Tennyson, following Whewell's thinking, manifests a uniformitarian poetics that realises textually the cultural consequences of the failure of teleological epistemologies.

Adamic and teleological discourses are supremely monologic. They pull towards a centre, closing down heteroglossia and attempting, as Trench and Whewell did, to lock the 'idea' and the 'word' into a hierarchical relationship in which the fixed idea is dominant. The idea is privileged in the same way as the monologic text privileges the author's worldview: "The authoritative word" Bakhtin writes, "is located in a distanced zone, organically connected with a past that is felt to be hierarchically higher. It is, so to speak, the word of the fathers. Its authority was already *acknowledged* in the past. It is a *prior* discourse."[24] While Trench and Whewell similarly take their authority from the past, uniformitarian thinking privileges instead the unlimited potential for change in the present. Language, seen in uniformitarian terms, is like the Lyellian landscape continuously shifting in the present, responding to laws that operate at a level hardly perceptible to individual consciousness, slowly transforming both words and their meanings. In *Maud*, the loss of teleological authority (the 'word of the fathers') begins with the literal loss of the father, which in turn "take[s] on the symbolic significance of the loss of all fathers" and all 'prior', authoritative discourses.[25]

It was not until the 1860s, nearly a decade after Tennyson's publication of *Maud*, that a paradigmatic shift in the approach to the historical-comparative study of language began, with the appropriation of uniformitarian methodologies by language theorists. Linguistics historian Craig Christy notes that if, as in Adamic theory, "development [in language] is seen as decay rather than as continuity or progress, it makes sense to focus linguistic analysis on the seemingly more perfect, or complete, system of older, written languages, rather than on the apparently imperfect, fragmentary testimony of spoken speech". However,

---

[24] M.M. Bakhtin, *The Dialogic Imagination*, ed. Michael Holquist, trans. Caryl Emerson and Michael Holquist (1981), 342. Hereafter cited parenthetically as *Dialogic*.

[25] Slinn, *Discourse of Self* (1991), 67.

uniformitarian thinking "led linguists to view not written language, but spoken speech as the most fruitful object of linguistic research".[26] A major figure in the appropriation of the uniformitarian methodology was the American philologist and lexicographer William Dwight Whitney, who wrote in 1867, "It is but a shallow philology, as it is a shallow geology, which explains past changes by catastrophe and cataclysms."[27] Whitney was "Philology's Lyell"; he was the "first to base upon this uniformitarian principle the principles of linguistic science".[28] "There can be no doubt" Christy asserts, "that Whitney was familiar, in particular, with Lyell's uniformitarian geology, as well as with the uniformitarian–catastrophist debate. This is evidenced by [Whitney's] references to the works of Lyell and Whewell."[29] Whitney saw modern language as a processive and mutable living force, and he "repeatedly stressed that language is a social institution, a conglomerate of innovating individuals, and that linguistic signs are, thus, but social conventions, subject to the circumstances and constraints of human will and need". Christy contends that Whitney's uniformitarian linguistics "inspired ... early structuralists" and suggests Saussure's indebtedness to his work.[30] Influenced also by Lyell and Whewell, Tennyson's own experimentation with language appears to pre-empt Whitney's thinking, producing in *Maud* a radical, anti-teleological and 'uniformitarian poetics' that is, as the next section hopes to show, dialogically orientated.

MAUD'S DIALOGISM

Maud, as Alan Sinfield has noted, is "problematic because it offers a powerful challenge to customary ideas of the relationship between author, text and reader".[31] This challenge, I argue below, is the result of the speaker's

---

[26] Craig Christy, *Uniformitarianism in Linguistics* (1983), 58.

[27] William Dwight Whitney, *Language and the Study of Language: Twelve Lectures on the Principle of Linguistic Science*, 1867 (1870), 287.

[28] Christy, *Uniformitarianism* (1983), x, 79. See also E.F. Konrad Koerner, "William Dwight Whitney and the influence of Geology on Linguistic Theory in the 19th Century," in *Language and Earth*, Bernd Naumann, Franz Plank and Gottfried Hofbauer eds. (1992), 271–287.

[29] Whitney, *Language* (1870), 79.

[30] Christy, *Uniformitarianism* (1983), 87, 85, 88.

[31] Alan Sinfield, *Alfred Tennyson* (New York: Blackwell, 1986), 168.

confrontation with language as dialogic, which is itself the consequence of seeing language through a uniformitarian lens that, as Whitney asserts in his later theories on language, shows it to be unfixed from the authority of the past and supremely protean in the present. Seemingly at odds with itself, the poem suffered at the hands of contemporary critics, who seemed to find themselves compromised by the text's indeterminacy. The poem appeared to be disconcertingly open; it churns out meaning endlessly, as contemporary criticism demonstrates. Robert James Mann, in his 1856 *Tennyson's 'Maud' Vindicated*, gives a flavour of the confusion that accompanied *Maud*'s reception. For one critic, he writes, the poem is "*a spasm*; another discovered it was a careless, visionary, and unreal *allegory of the Russian war*; a third could not quite make up his mind whether the adjective mud or mad would best apply to the work ... A fourth found that the mud concealed irony". Critics read *Maud* as "a *political fever*,—an *epidemic* caught from the prevalent carelessness of thought and rambling contemplativeness of the time;—*obscurity mistaken for profundity*,—the dead level of prose run mad" and so on.[32] Such indeterminacy is a mark of *Maud*'s dialogic engagement, which is extensive and complex.

For example, *Maud* foregrounds the utterance on many levels, it figures a speaker whose tenuous sense of self is overwhelmed by his confrontation with heteroglossia—a speaker whose mentality is monologic but whose consciousness is given in the text as dialogic, immediate and unfinalisable. This is the extreme manifestation of the same dialogic condition experienced by *In Memoriam*'s speaker. Where the consequences of losing monologic unity are dire for the speaker of that poem—so close to the author—*Maud*'s speaker can more safely follow the trajectory of uniformitarianism's way of seeing. Inevitably, as in *In Memoriam*, the depiction of a dialogic consciousness creates problems for textual closure. This is so because, like consciousness itself, the textual representation of consciousness cannot enact its own closure. Just as *In Memoriam* had to introduce a teleological narrative in order to transcend the poem's uniformitarian form, Part III of *Maud* addresses the problem of closure by shifting from a dialogic to a monologic perception of language. This section reads Parts I and II of *Maud* in conjunction with Bakhtinian theory, examining Tennyson's creation of a dialogic consciousness. Finally, the turn from dialogism to monologism in the poem's final stage is

---

[32] Robert James Mann, *Tennyson's 'Maud' Vindicated: An Explanatory Essay* (1856), 6–7.

examined, as this shift is fittingly accompanied by a switch from a uniformitarian metaphoric structure to one shaped by catastrophism.

Tennyson's dialogic achievement in *Maud* is best understood through Bakhtin's analysis of Dostoevsky and the polyphonic novel. *Maud*, however, is not polyphonic in the way Bakhtin suggests novels can be. Polyphony describes a specific artistic form that for Bakhtin could only be fully realised in the novel: the polyphonic novel is one in which a *"plurality of independent and unmerged voices and consciousnesses"* are given within the bounds of the artistic work. *Maud*, however, cannot and clearly does not set out to do this.[33] Nevertheless, Bakhtin's account of Dostoevsky's depiction of consciousness has a strong bearing on Tennyson's creation of *Maud*'s speaker. "What is important to Dostoevsky", Bakhtin writes, "is not how his hero appears in the world but first and foremost how the world appears to his hero, and how the hero appears to himself" (*Problems*, 47). The speaker of *Maud* is continually interpreting and reinterpreting himself to himself, offering the reader his unknowable self, his unfinalisable consciousness as he experiences it and not the author's circumscribed creation of it. His monologue is therefore not a psychological snapshot from which the reader may extrapolate further, as in those dramatic monologues that encourage readers to make psychological judgements, as this would be to give the reader what Bakhtin would consider a false sense of the knowability or the finalisability of consciousness (*Problems*, 61–62). Such a monologue offers the reader the author's or the 'other's' view of the speaking 'I', and not the self's view of the self. Rather, *Maud*'s speaker is awkwardly unfinalisable and resists psychologising; he is only discernible to readers in the event of the microdialogue of consciousness. Thus, Tennyson creates in his speaker a "person", not a "character" in the sense in which Bakhtin distinguishes between "person" and "character". As Michael Holquist explains: "Character is a monologic, finished-off, generalized category that is given and determined ... Person, on the other hand, is a dialogic, still-unfolding, unique event."[34] It is this confrontation with a dialogic consciousness that disoriented those contemporary readers who came to Tennyson's monodrama anticipating the monologic vision of the author.

---

[33] Mikhail Bakhtin, *Problems of Dostoevsky's Poetics*, Carl Emerson ed. and trans. (1984), 6. Hereafter cited parenthetically as *Problems*.

[34] Michael Holquist, *Dialogism: Bakhtin and his World* (1990), 162.

Tennyson achieves the textual representation of consciousness through a relinquishing of his "surplus of seeing"—his omniscient knowledge of his fictional world—a condition essential, Bakhtin argues, for the artistic creation of a polyphonic text. Bakhtin elucidates: "Everything that the author-monologist kept for himself, using it to create the ultimate unity of a work and the world portrayed in it [is ...] turn[ed] over to his hero, transforming all of it into an aspect of the hero's self-consciousness". In this way, the author "constructs not a character, not a type, nor a temperament. In fact he constructs no objectified image of the hero at all, but rather the hero's *discourse* about himself and his world" (*Problems*, 52, 53). Tennyson relinquishes his "surplus of seeing" to depict a speaker who experiences himself as unfinished and ongoing in a provisional and ever-changing present. Such immediacy and indeterminacy results from Tennyson's experimental uniformitarian poetics that prioritise present processes over past events, thus allowing the poet to treat the subject in language as processive, dynamic and liberated from authorial prefixed determinates. Just as *Maud*'s uniformitarian landscape shifts endlessly, so the speaker's sense of self is also subject to continual process. As a result, the text is able to depict fluid consciousness rather than fixed character.

The creation of a dialogic consciousness has a deeply political resonance. Bakhtin discusses how the dialogic author re-creates the complexity of the "human soul" in order to counter a reductive psychology that sees individual consciousness as finalisable and knowable. The "major emotional thrust of all [Dostoevsky's] work, in its form as well as its *content*, is the struggle against a *reification* of man, of human relations, of all human values under the conditions of capitalism" (*Problems*, 62). In a similar vein, *Maud* posits an authentic, unknowable and unfinalisable consciousness against the "hubbub of the market" (II, 208) and the "mammonism" of industrial economics. In *Maud*'s uniformitarian world, where deep time is accelerated in order to manifest those geological processes that petrify the living, the speaker's consciousness becomes the last refuge of human values, a last defence against the stolid and stultifying progress of capitalism. Thus, Tennyson apprehends, similarly to Dostoevsky, how the "*reifying devaluation* of man had permeated into all the pores of contemporary life, and even into the very foundations of human thinking", and he posits against this reification a fluid, dialogic consciousness that resists hypostatisation through its claim to be present, ongoing, unstable and empirically unknowable (*Problems*, 62).

In keeping with the dialogic turn, the author's relinquishing of his "surplus of seeing" allows voice and hearing to become prioritised.[35] While *Maud* has only one speaker, this speaker's consciousness crosses, intersects and responds to a multitude of other voices. In fact, everything seems to have a voice, from the "crying and calling" (I, 415) of the birds, the voices of the flowers in Maud's garden, the "dry-tongued laurel's pattering talk" (I, 606), to the "scream of a maddened beach" (I, 98). The human voice is figured in the "idiot gabble", the "crowd confused and loud" (II, 211), Maud's singing voice, the speaker's voice, but perhaps most tellingly, in the importance Tennyson placed on the recital of *Maud* as a means of allowing others to fully comprehend the poem. Famously, *Maud* was the poem Tennyson "loved best to read aloud and the one that he read most often and most powerfully; it was, above all, the one he most wished others to feel and understand" (*Memoir*, I, 395–6). Tennyson's son described his father's "last reading of 'Maud' ... His voice, low and calm in everyday life, capable of delicate and manifold inflection, but with 'organ-tones' of great power and range, [which] thoroughly brought out the drama of the poem".[36] Gladstone and the critic Henry Van Dyke, Tennyson's son reports, "wholly misapprehended the meaning of 'Maud' until they heard my father read it" after which they "publicly recanted their first criticisms" (*Memoir*, I: 398). Van Dyke confessed that he "misunderstood *Maud* and underrated it", but was "enlightened by hearing the poet read the poem aloud".[37] To Charles Richard Weld, Tennyson wrote, *Maud* "is an *entirely new form*, as far as I know. I think that properly to appreciate it you ought to hear the author read it" (*Letters ALT*, II, 135). And, when the essayist George Brimley suggested that the speaker of *Maud* is "the writer of the fragments of a life which tell the story of *Maud*", Tennyson replied saying, "I should take exception to...'the writer of the fragments, etc.,' surely the speaker or the thinker rather than the writer" (*Memoir*, I: 408).[38] Tennyson, it seems, felt *Maud* to be a verbal realisation of being; a manifestation of consciousness in its relationship with the world. This heavy investment in the poem's recital points to how *Maud* is both uniformitarian and

---

[35] For an extensive discussion of the centrality of the voice for Tennyson's understanding of language, see Donald S. Hair, *Tennyson's Language* (1991), 58–74.

[36] Henry Van Dyke, *Studies in Tennyson* (1920), 94.

[37] Ibid., 121–2.

[38] George Brimley (1855), 227

dialogic. It is uniformitarian in the sense that, as Whitney found, the uniformitarian approach to language shifts emphasis from the written to the spoken word, and dialogic in terms of Tennyson's awareness that the 'real' life of *Maud* is to be found in the relations between speaker and addressee, in the utterance itself, and in the intonation that registers the other's presence and draws the listener into dialogic responsiveness.

Critical responses also point to how dialogic relations between speaker and addressee are not static but movable through time and space. Dialogism "assumes" that the "text is always in production", there is no transhistorical interpretation, only a set of ever-changing relations between speaker, addressee and the world. With each reading meaning is remade because relations between speaker and addressee are remade at a new and unique point in time and space, and this "emphasis on the text's groundedness in a social and historical context *at every point of its existence* is one of dialogism's distinctive features".[39] Bakhtin describes this as the "chronotopic situation" of the listener/reader; "his role in *renewing* the work of art (his role in the process of the work's life)" (*Dialogic*, 257). The listener/reader's "chronotopic" position describes a set of ever-changing time/space co-ordinates that interact with the text's own chronotopes (with the social and historical time/space coordinates of the text's creation) and the "temporal and spatial relationships that are artistically expressed" in the text (*Dialogic*, 84). These infinitely variable permutations are what Holquist describes as the "multiplicity of chronotopes" produced by a single literary text.[40] Tennyson's compulsion to recite *Maud* expresses a desire to repeatedly resituate and refigure the poem's chronotopic relations, to continually configure new sets of temporal/spatial relations in a way that implies the author's awareness of the potential for the speaker's dialogue to be continually re-invested with meaning beyond the limits of the time and space of the poem's conception.

In *Maud*'s second section, the emphasis is on sound and the power of the utterance in the unofficial, unregulated social milieu. However, it also offers a specific critique of the politics of liberal reform—one that is also shaped by uniformitarian principles. Part II registers the way uniformitarian methodologies shift linguistics towards a radically democratising perception of language, empowering the users of language and legitimising the speaker. Buried beneath the city streets, the speaker hears

---

[39] Holquist, *Dialogism* (1990), 141.

[40] Ibid., 140.

the "Clamour and rumble, and ringing and clatter" (II, 251) of urban agitation. A characteristically Carlylean din that manifests what had been in the 1830s and 1840s the Chartist plea for suffrage—the "clattering of ballot-boxes" and the "infinite sorrowful jangle" produced by the mix of "inarticulate" voices and the mechanical din of an urban and industrial age.[41] These voices of the streets, however, need to be distinguished from the specific voices of the "lord", the "statesman" and the "physician" (II, 270–274). These voices represent the landed and professional classes, who are "sobbing", "blabbing", "betraying" and speaking "nothing but idiot gabble" (II, 268–274, 279). For the speaker, they represent the "idiocy" of liberal reform, not the "sullen thunder" of the disenfranchised but the insane "gabble" of the enfranchised themselves: those who the speaker identifies as of his own class, as "one of us" (II, 268). The "idiot gabble" depicts a "babbling" democracy that does not "let any man think for the public good" (II, 283); a Babel in which there is no one "man and leader of men" (I, 388), no ruling, principal voice; a Babel in which all speak but none listen, and by speaking all at once, all is made unintelligible. Thus, the liberal impulse towards democratic reform is experienced as an overwhelming heteroglossia that entirely subsumes authoritative discourses. Part II culminates in the speaker's life-in-death experience, the source of which is the realisation that, in an anti-teleological, uniformitarian landscape, language is no longer rooted in Adamic origins—no longer fixed as in fossil poetry—but instead shifts endlessly according to the vicissitudes of social contexts. To the speaker's monologically orientated mentality, this view of language is experienced as a type of madness. Herbert Tucker points out that "Etymologically *idiocy* is privacy". Thus the "idiot gabble" can be also be read as representing the potency of the uncensored, private utterance to effect linguistic change and override Adamic prefixed meanings in language.[42] What is destabilising is the realisation that speech dissolves monologism, that the individual's utterance is the dialogic engine of ideological/social change; in other words, that meaning is made and remade continually in the "babbling" exchange of social discourse.

This is Whitney's view of language that he arrives at through his extrapolation of Lyell's uniformitarian methodology across the field of linguistics. Where Lyell cites processes of gradual erosion, sedimentation

---

[41] Thomas Carlyle, *Chartism* 1839 (1840), 53.

[42] Tucker, *Doom of Romanticism* (1988), 426.

and uplift as ubiquitous yet imperceptible agents of the profoundest geological change, so Whitney sees language as "undergoing, everywhere and always, a slow process of modification, which in the course of time effects a considerable change in its constitution, rendering it to all intents and purposes a new tongue".[43] Confronted by a heteroglossic "idiot gabble", the speaker's instinct is to recoil, but the desire to bury himself in himself only leads to the alarming awareness that consciousness itself is dialogic and reliant for its existence in language on the social nexus (a perception that the author of *Maud* acknowledges in his own urge to recite continually the poem as a means of generating meaning). The "shallow grave" offers no peace and buried beneath the city streets, at the dialogic crossroads, the life-in-death experience becomes a graphic aesthetic enactment of the dialogic truism that "self-consciousness cannot be finalized *from within*" (*Problems*, 73). Like the anti-teleological uniformitarian world in which there is "no vestige of a beginning, no prospect of an end", the dialogic 'I' is processive and "constantly open"; it "has no beginning and no end".[44] Thus, the madhouse canto's aesthetic representation of heteroglossia registers not only specific fears of an ideological breakdown wrought by a babbling liberal democracy but also the fear of relinquishing the Adamic belief (and also *In Memoriam*'s crowning hope) that consciousness will ultimately be remerged into the original 'authoritative word'—subsumed into a supreme transcendental signifier. This is a hope that *Maud*'s uniformitarian, post-teleological perception of being finally debunks.

As a "subversive reactionary" whose "conservatism" nevertheless "leads him towards radical questioning", Tennyson shares with Dostoevsky a deep rooted political ambivalence.[45] Bakhtin describes Dostoevsky as "oscillat[ing] between a revolutionary materialistic socialism and a conservative religious worldview—oscillations that never led to any decisive resolution". Such irresolvable contradictoriness is the key to Dostoevsky's creation of polyphony (*Problems*, 34). The polyphonic text demands of the author an "extraordinary artistic capacity for seeing everything in coexistence and interaction". The polyphonic author hears in every voice "two contending voices, in every expression a crack, and the

---

[43] Whitney, *Language* (1870), 34.

[44] Holquist, *Dialogism* (1990), 37.

[45] Isobel Armstrong, *Victorian Poetry* (1993), 2, 283.

readiness to go over immediately to another contradictory expression". In Dostoevsky's novels, these "contradictions and bifurcations... were never set in motion along a temporal path or in an evolving sequence: they were, rather, spread out in one plane, as standing alongside or opposite one another, as consonant but not merging or as hopelessly contradictory, as an eternal harmony of unmerged voices or as their unceasing and irreconcilable quarrel" (*Problems*, 30). Such immediacy also closely resembles the uniformitarian state of being. Just as the polyphonic author must see everything in coexistence and unceasing interaction, Lyell subtly configures a processive world in which every cross-section of every moment reveals the unending contradiction in which geologic agents are simultaneously antagonistic and harmonious. With no recognition of first or last causes, the uniformitarian eye sees only what is occurring now, and sees what is occurring now as in an interminable state of becoming. There can be no directional or evolutionary movement in Lyell's system, as the uniformitarian world exists in any given moment poised between ruin and renewal, perpetually forming and re-forming itself in processes of displacement in which there is no purpose or gain. Lyell writes, "The renovating as well as the destroying causes are unceasingly at work, the repair of the land being as constant as its decay ... If the separate effects of different agents, whether aqueous or igneous, are insensible, it is because they are continually counteracted by each other" (*PG*, I, 473–4). Tennyson's dialogism is the result of his understanding of uniformitarian principles, his willingness to experiment with and to interrogate Lyell's and Whewell's ostensibly scientific discourses, as well as his own intrinsic ambivalence, all of which lead to an irresolvable contradictoriness that finds its aesthetic expression in the speaker's dialogic sensibility.

The final section of *Maud* is notoriously problematic. As Christopher Johnson notes, it "seems to be on the one hand the only possible resolution, yet on the other no resolution at all, but a self-sacrifice full of irony", a sacrifice that is neither "satisfactory [n]or wholly convincing".[46] The problem stems from the speaker's inability to finalise himself. His imagined death suggests the urge to make whole the fractured self, and in doing so, to terminate the text's dialogism. Dislodged from the "high Hall", the speaker descends to street level, where he is subsumed into the geological strata. From here, his perception of himself as moving between the linear co-ordinates of a fixed past and a prefigured future

---

[46] Christopher Johnson, "Speech and Violence in Tennyson's *Maud*" (1997), 57–58.

dissolves, revealing a borderless uniformitarian infinity that offers no peace but instead continually defers and denies completion in its unending dialogic exchange. Only in the recovery of a monologic perception of the world can the illusion of wholeness and finalisability be restored. This represents another of *Maud*'s ironies; Tennyson's astonishing creation of a dialogic consciousness dictates that the poem's closure must inevitably undermine his brilliant aesthetic achievement.

The shift in the speaker's register in Part III indicates the closing down of dialogism. Bakhtin describes, again with reference to Dostoevsky's polyphonic novels, how "An enormous and intense dialogic activity is demanded of the author of a polyphonic novel: as soon as this activity slackens, the characters begin to congeal, they become mere things, and monologically formed chunks of life appear" (*Problems*, 68). Only with the emergence of an authorial narrative voice can the hero's consciousness be reined in, allowing the closure (that consciousness cannot enact itself) to take place. In Part III, the speaking voice loses its dialogic fluidity, fixing itself instead around a set of immutable ideologies articulated through the trite rhetoric of the "defence of the right" and the "glory of manhood" (III, 19, 21). Such rhetoric recalls Maud's war song, which appears as a "monologic chunk", or, for the speaker at least, a monologic anchor amid the dialogic uncertainty of Part I. It represents a fixed ideological point of reference, teleologically prefiguring the poem's final section and the "doom assigned". Maud's song is the transhistorical aesthetic object—the ostensibly unanswerable monologic voice of the past, and it is this monologic wholeness that Maud represents and that is the central object of desire in the poem. In Part III, authorial control is asserted through the speaker's submission to the fixed rhetoric of Maud's song and through the author's investment in *Maud* as a finished aesthetic object. The introduction of a monologic register is designed to reinstate an ideological centre and to correct the madness—the audio defect of heteroglossia—that is *Maud*'s "idiot gabble". The awkwardness of this shift, however, is palpable, as so many readers have noted.

The movement from dialogism to monologism is accompanied by a shift away from uniformitarian tropology. The slow and creeping petrifaction that foregrounds insidious processes of geological change, and through which reification is figured, is replaced by images of war and sudden catastrophe. Catastrophe is brought into relief by the resumption of human time. In human time, catastrophes may happen with devastating

local effect—catastrophe may mean death for the speaker—but this is nevertheless a restoration of normal human experience and a release from a
humanly alien uniformitarian view of change in which catastrophe appears
ineffectual. An erosive "peace" represented by the "cannon-bullet" that
"rusts on a slothful shore" (III, 26) is replaced by the "clash" of war (III,
44) in which "many a light shall darken", and "many a darkness into the
light shall leap" (III, 43, 46). Gradual uniformitarian change is replaced
by "sudden" catastrophic reversals. Fittingly, those who fall are "crushed
in the clash of jarring claims" (III, 44) in a line that combines the notion
of land as disputed property with images of the "clash" and "jarring"
of the land in the geographical catastrophe of the earthquake. From
the "shallow grave", the speaker, along with the "glory of manhood",
is restored to an "ancient height" (III, 21) in an action that represents
the reactionary re-emergence of a buried past and the resurrection of the
authority of the 'word of the father'. The speaker has "awaked ... to a
better mind" (III, 56) in which his paralysis is dispelled by his absorption
into what is an ideal hierarchical form suggested in the pyramidic structure of the military. Here, order imposed from above ensures that eve-
ryone speaks the same language and, fittingly, a "loyal people" shout in
unison "a battle cry" (III, 35) that converts *Maud*'s dialogic "babble"
into monologic sense. The "higher" authority of an ideological frame signifies a return to a teleological mentality that foresees final causes in "the
purpose of God, and the doom assigned" (III, 59). Thus, the finalisation
of consciousness is achieved as the speaker is re-integrated into the 'prior
discourse' of Maud's war song, and the author's omniscient vision.

War also suggests for the speaker, who is "nameless and poor" (I,
1196), the possibility of gaining a name in death in the "Honour that
cannot die" (I, 177). Christopher Johnson has noticed the "pathos
in the speaker's hope that 'many a darkness' may 'shine in the sudden
making of splendid names' [as] The seashell had not needed a 'clumsy
name'". The speaker's sanity "is undermined by his own abrupt clarity,
his sudden and disturbing faith in words and names".[47] The return to a
faith in names is a return to a Whewellian and Trenchian privileging of
the idea over the word. The transmutive uniformitarian vision of the sliding scale of people and things that linguistically renders "names unmeaning", is here arrested by a return to a faith in the fixity of types and the

---

[47] Johnson, "Speech and Violence" (1997), 60.

integrity of names. In this monologic world, it is not Maud the living woman who signifies but the fixed *idea* of "war" and "battle" that pre-exists etymologically in her name. Maud becomes the "transcendental signified" that ostensibly allows the speaker to rise above the continually shifting materiality of the poem's landscape and language into the rarefied realm of the idea.[48] Here, it is not the living speaker's words or consciousness that generate meaning but dead men whose "splendid names" represent those fixed ideas for which living men must die. The return to teleology offers some sense of poetic closure, while the speaker's re-integration into the authorial frame closes down the anarchic consciousness that pervades the earlier sections of the poem. Yet, despite Part III's monologism, and despite the attempt to refigure *Maud's* uniformitarian imagery through catastrophe, the poem's dismantling of teleological epistemologies is so complete that rebuilding them convincingly becomes impossible. Thus, the final section, rather than enacting a satisfactory closure, merely fulfils the speaker's need to be re-subsumed into a monoglossic order that the poem itself leads readers to distrust.

Maud performs Lyell's uniformitarian principles, restructuring in the process the epistemological basis of language. In its bold experimentation with a uniformitarian poetics, it not only manifests dialogic uncertainty by relinquishing the fixed categories upon which Trench and Whewell insist, it also pre-empts advances in the science of linguistics that would later lead to a perception of words as floating social signs detached from any intrinsic relationship with that which they signify. Read in this light, *Maud's* "babble" and "gabble" is a heteroglossic disorder that dissolves the illusion of monologic cohesion and undermines not only the authorial voice but also that highest authority presupposed by Adamic theory. It is no wonder that Tennyson regarded *Maud* as exceptional, that he could never be finished with it, that he felt the need to re-animate its utterance again and again, re-entering continually the field of *Maud's* interminable dialogue. The poem's remarkable experimentation produces a uniformitarian poetics characterised by its dialogism; it also offers critical evidence of the profound cultural and linguistic impact of the uniformitarian methodology on pre-Darwinian sensibilities.

---

[48] Sinfield, *Alfred Tennyson* (1986), 170.

Afterword: What Remains

Charles Darwin famously said with reference to Lyell's *Principles of Geology*, "The great merit of the *Principles* was that it altered the whole tone of one's mind, and therefore that when seeing a thing never seen by Lyell, one yet saw it partially through his eyes."[1] Tennyson's poetry suggests that the same was true for the poet. His mid-century poems are remarkable discourses that, as I have attempted to argue, are geological texts in themselves. They are so because Tennyson, like the rest of his generation, saw literature and science as expressive of the same fundamental truths. He felt the power of language as a reality, but he also felt that power ebbing away; the shift from Tennyson's confident use of Hugh Miller's geology in *The Princess* to the Lyellian crises of *In Memoriam* and *Maud* charts a movement from the centre—an Adamic point of meaning—to the periphery—the very edge of meaning itself. Tennyson's poetry traced an epistemological shift that led him to an intimate examination of his faith, not only in religious or spiritual terms but in terms of his faith in his own art, in language and its ability to embody the 'truths' of geology, or, in fact, any other truths. The poems anticipate the later crisis that led to the fragmentation of knowledge—a crisis in which Lyell's *Principles* had a vital role to play. Tacitly

---

[1]Charles Darwin to Leonard Horner, Aug 29, 1844, Burkhardt ed., *Correspondence* (1987), III, 55.

© The Editor(s) (if applicable) and The Author(s) 2017
M. Geric, *Tennyson and Geology*, Palgrave Studies in Literature, Science and Medicine, DOI 10.1007/978-3-319-66110-0

apprehending the import of Lyell's uniformitarianism, Tennyson produced a 'uniformitarian' poetics that was astonishing for how it materialised Lyell's way of seeing, and in the process, prefigured modernity's perception of language.

While this book has attempted to demonstrate some of the issues that emerge from Tennyson's mid-century poetic experimentation with geology, much has been left unexamined, and the significance of geology more generally for the first half of the nineteenth century has yet to be explored in depth. Thus, one of the broadest aims of the book has been to suggest that the moment just before Darwin's theory of evolution exploded on the nineteenth-century world, a moment when geology was at the height of its prominence, is a much neglected area of criticism. William Dyce's evocative painting, *Pegwell Bay: A Recollection of October 5th, 1858* (1858–1860), provides a brilliant visual depiction of Lyellian deep time and hints at what has been overlooked. The figures of Dyce's painting appear oddly disengaged from the setting, and in this they aptly express the Lyellian geological perception of the human condition—one in which individuals, subject to Lyell's 'strategy of division', find themselves alienated from nature and disconnected from the material world by the crushing reality of geological time. Dyce's painting epitomises the sense of detachment from nature and of distance from that which science studies. And it is a painting that is particularly resonant because it specifically depicts that moment just before Darwin's publication of *On the Origin of Species* (1859)—just before, in other words, evolutionary theory took centre stage as the most potent and controversial scientific idea of the nineteenth century. Whether evolution was welcomed or resisted, what Darwin's theory did was to put humanity back into nature, re-situating human beings in a line of descent that suggested somehow, and in some form, they had always been at the centre of planetary life. The event of Darwin's publication resulted, however, in the overshadowing of the very specific and singular effect of geology and particularly of Lyell's *Principles* on the first half of the nineteenth century, and of other fascinating practices that came under the banner of geology in the period, such as comparative anatomy (Gowan Dawson's work is crucial here). The decades before Darwin are highly significant not only for understanding Tennyson and geology but also for the history of ideas, for thinking about the relationship between literature and science, and for understanding the development of modern critical thinking. Geology's paradigmatic moment has not been fully examined and deserves a much closer exegesis, as its significance did not disappear with the rise of Darwinian theory. Rather, the polemics of this protean period had, and continue to have, a significant influence on modern thinking. This book has read that polemic in Tennyson's mid-century long poems—texts that are themselves significant geological discourses—arguing that the poems demonstrate how poetry and geology, literature and science begin to pull apart and come to be seen as different kinds of knowledge subject to relative values. It has attempted in part to recover something of the pre-Darwinian, Lyellian shift from unity to division and from a sense of centredness to that of alienation, a moment depicted in *Pegwell Bay*, and one which Tennyson's poetry so profoundly expresses in its uniformitarian poetics.

BIBLIOGRAPHY

Aarsleff, Hans. *From Locke to Saussure: Essays on the Study of Language and Intellectual History.* London: Athlone, 1982.

———. *The Study of Language in England 1780–1860.* Minneapolis: University of Minneapolis Press, 1983.

Adams, James Eli. "Woman Red in Tooth and Claw: Nature and the Feminine in Tennyson and Darwin," in *Tennyson,* edited by Rebecca Stott. London, New York: Longman, 1996.

Anderson, Lyall I., and Michael A. Taylor. "Tennyson and the Geologists Part 1: the Early Years and Charles Peach," *The Tennyson Research Bulletin,* 10, no. 4 (2015): 348–350.

Armstrong, Isobel. *Language as Living Form in Nineteenth Century Poetry.* Sussex: Harvester Press, 1982.

———. *Victorian Poetry: Poetry, Politics and Poetics.* London: Routledge, 1993.

Bailin, Miriam. *The Sick Room in Victorian Fiction: The Art of Being Ill.* Cambridge and New York: Cambridge University Press, 1994.

Bakhtin, Mikhail. *The Dialogic Imagination: Four Essays by M. Bakhtin.* Edited by Michael Holquist, translated by Caryl Emerson and Michael Holquist. Austin, Texas: University of Texas Press, 1981.

———. *Problems of Dostoevsky's Poetics.* Edited and translated by Caryl Emerson. Manchester: Manchester University Press, 1984.

Bartholomew, Michael, "Lyell and Evolution: An Account of Lyell's Response to the Prospect of an Evolutionary Ancestry for Man." *British Journal for the History of Science* 6 (1972–1973): 261–303.

———. "The Non-Progress of Non-Progression: Two Responses to Lyell's Doctrine." *British Journal for the History of Science* 9 (1976): 166–174.

© The Editor(s) (if applicable) and The Author(s) 2017

M. Geric, *Tennyson and Geology,* Palgrave Studies in Literature, Science and Medicine, DOI 10.1007/978-3-319-66110-0

Barton, Anna. *Tennyson's Name: Identity and Responsibility in the Poetry of Alfred Lord Tennyson.* Birmingham: Ashgate, 2008.

Beer, Gillian. *Darwin's Plots: Evolutionary Narrative in Darwin, George Eliot and Nineteenth-Century Fiction.* 1983. Cambridge: Cambridge University Press, 2000.

———. *Open Fields: Science in Cultural Encounter.* Oxford: Oxford University Press, 1996.

Beesemyer, Irene Basey, "'Black with the void from which God himself has disappeared': Spatial Displacement in Tennyson's Maud." *Tennyson Research Bulletin* 7, no. 4 (2000): 174–192.

Bennett, William Cox. *Anti-Maud.* London: E. Churton, 1855.

Brimley, George. "Alfred Tennyson's Poems." In *Cambridge Essays.* London: John W. Parker, 1855.

Bristow, Joseph. "Nation, Class, and Gender: Tennyson's *Maud* and War." *Genders* 9 (1990): 93–111.

Brooke, Stopford A. *Tennyson.* London: Isbister, 1898.

Brown, Daniel. *The Poetry of Victorian Scientists: Style, Science and Nonsense.* Cambridge: Cambridge University Press, 2013.

Buckland, Adelene. *Novel Science.* Chicago and London: University of Chicago Press, 2013.

Buckland, William. *Geology and Minerology Considered with Reference to Natural Theology,* 2 vols. London: Pickering, 1836.

Burke, Edmund. *Reflection on the Revolution in France,* edited by L. G. Mitchell. Oxford: Oxford University Press, 1993.

Burkhardt, Fredrick, and Smith, Sydney, eds. *Correspondence of Charles Darwin.* Cambridge: Cambridge University Press, 1987.

Burnet, Thomas, *Sacred Theory of the Earth.* London: 1691. Facsimile reprint of 2nd edition. London: Centaur Press, 1965.

Campbell, Nancie, ed., *Tennyson in Lincoln: A Catalogue of the Collections in the Research Centre.* Lincoln: Tennyson Research Society, 1971.

Carlyle, Thomas. *Chartism.* 1839. London: James Frazer, 1840.

———. *Past and Present.* London: Chapman and Hall, 1843.

———. *Latter-Day Pamphlets.* London: Chapman and Hall, 1850.

———. *Historical Sketches of Notable Persons and Events in the Reigns of James I and Charles I.* Ed. Alexander Carlyle. London: Chapman and Hall, 1898.

Chambers, Robert. *Vestiges of the Natural History of Creation.* 1844. Facsimile reprint of 1st edition. Chicago: University of Chicago Press, 1994.

Christy, T. Craig. *Uniformitarianism in Linguistics.* Amsterdam/Philadelphia: John Benjamins Publishing, 1983.

Clark, Katerina, and Michael Holquist. *Mikhail Bakhtin.* Cambridge, MA: Harvard University Press, 1984.

Collins, Philip, ed., *Tennyson: Seven Essays.* Basingstoke: Macmillan, 1992.

Corsi, Pietro. "The Importance of French Transformist Ideas for the Second Volume of Lyell's Principles of Geology." *British Journal for the History of Science* 10–11 (1977–1978): 221–244.

Cregan-Reid, Vybarr. *Discovering Gilgamesh*. Manchester: Manchester University Press, 2013.

Crowley, Tony, *The Politics of Discourse: The Standard Language Question in British Cultural Debates*. London: Macmillan, 1989.

Culler, A. Dwight. *The Poetry of Tennyson*. New Haven and London: Yale University Press, 1977.

Curran Stuart. *Poetic Form and British Romanticism*. Oxford: Oxford University Press, 1986.

Darwin, Charles. *The Voyage of the Beagle*. Hertfordshire: Wordsworth Editions, 1997.

Dawson, Gowan. *Reconstructing Prehistoric Monsters in Nineteenth-Century Britain and America*. University of Chicago Press, 2016.

———. "Dickens, Dinosaurs, and Design." *Victorian Literature and Culture* 44 (2016): 761–778.

———. "Literary Megatheriums and Loose Baggy Monsters: Paleontology and the Victorian Novel." *Victorian Studies*, 53, no. 2 (2011): 203–230.

Dawson, Samuel Edward. *A Study of Lord Tennyson's Poem, The Princess*. Montreal: Dawson Brothers, 1882.

Day, Aidan. *Tennyson's Scepticism*. Basingstoke and New York: Palgrave Macmillan, 2005.

Dean, Dennis R. "Geology and the Victorians." In *Victorian Science and Victorian Values: Literary Perspectives*, edited by J. Paradis and Thomas Postlewait. New York: New York Academy of Sciences, 1981.

———. *Tennyson and Geology*. Lincoln: Tennyson Research Centre, 1985.

Desmond, Adrian. "Artisan Resistance and Evolution in Britain, 1819–1848." *Osiris* 3 (1987): 77–110.

———. *The Politics of Evolution: Morphology, Medicine, and Reform in Radical London*. Chicago and London: University of Chicago Press, 1992.

Douglas-Fairhurst, Robert. *Victorian Afterlives: The Shaping of Influence in Nineteenth-Century Literature*. Oxford: Oxford University Press, 2002.

Duncan, Henry. "An Account of the Tracks and Footmarks of Animals found impressed on Sandstone in the Quarry of Corncockle Muir, in Dumfriesshire." *Transactions of the Royal Society of Edinburgh* XI, no. 11 (1828): 194–209.

Eliot, T. S. *Essays Ancient and Modern*. New York: Hardcourt, 1936.

Eliot, T. S. "In Memoriam," in *Critical Essays on the Poetry of Tennyson*, edited by John Killham. London: Routledge & Kegan Paul, 1964.

Gates, Sarah. "Poetics, Metaphysics, Genre: The Stanza Form of *In Memoriam*." *Victorian Poetry* 37 (1999): 507–520.

Geric, Michelle. "Shelley's 'cancelled cycles': Huttonian Geomorphology and Catastrophe in *Prometheus Unbound*." *Romanticism* 19, no. 1 (2013): 31–43.

Gibson, Walker. "Behind the Veil: A Distinction Between Poetic and Scientific Language in Tennyson, Lyell, and Darwin." *Victorian Studies* 2 (1958): 60–68.

Gliserman, Susan. "Early Victorian Science Writers and Tennyson's "In Memoriam": A Study in Cultural Exchange." *Victorian Studies* 18 (1975): 277–308.

Gould, Stephen Jay. *Time's Arrow, Time's Cycle: Myth and Metaphor in the Discovery of Geological Time*. London: Penguin Books, 1987.

Grierson, James. "On Footsteps Before the Flood in a Specimen of Red Sandstone." *Edinburgh Journal of Science*, Art. XXI (1828): 130–134.

Griffiths, Devin. *The Age of Analogy: Science and Literature Between the Darwins*. Baltimore, MD: John Hopkins University Press, 2016.

Griffiths, Eric. "Tennyson's Idle Tears," in *Tennyson Seven Essays*, edited by Philip Collins. London: Palgrave Macmillan, 1992.

Hair, Donald S. *Tennyson's Language*. Toronto: University of Toronto Press, 1991.

———. ""Soul and Spirit" in *In Memoriam*." *Victorian Poetry* 34 (1996): 175–191.

Hallam, Arthur. *Remains in Verse and Prose of Arthur Henry Hallam*. London: W. Nicol, 1834.

Hamilton, William, *Observations on Mount Vesuvius, Mount Etna, and other Volcanoes*. London: T. Cadell, 1774.

Henchman, Anna, ""The Globe we groan in": Astronomical Distance and Stellar Decay in *In Memoriam*." *Victorian Poetry* 41, no. 1 (2003): 29–45.

Heringman, Noah. *Romantic Rocks, Aesthetic Geology*. Ithaca and London: Cornell University Press, 2004.

Holmes, John. *Darwin's Bards: British and American Poetry in the Age of Evolution*. Edinburgh: Edinburgh University Press, 2009.

Holquist, Michael. *Dialogism: Bakhtin and his World*. New York: Routledge, 1990.

Hutton, James. *The Theory of the Earth*. 3 vols. Edinburgh: Cadell, Jr. and Davies, 1795. Facsimile edition. Caldicote, Herts: Engelmann, Whelcon and Wesley, 1959.

Johnson, Christopher. "Speech and Violence in Tennyson's Maud." *Essays in Criticism* 47 (1997): 33–61.

Kendall, J. L. "Gem Imagery in Tennyson's Maud." *Victorian Poetry* 17 (1979): 389–394.

Killham, John. *Tennyson and The Princess: Reflections of an Age*. London: University of London Press, 1958.

————. "Tennyson's *Maud*. The Function of the Imagery." In *Critical Essays on the Poetry of Tennyson*, edited by John Killham. London: Routledge & Kegan Paul, 1964.

Klaver, J. M. I. *Geology and Religious Sentiment: The Effect of Geological Discoveries on English Society and Literature between 1829 and 1859*. New York: Brill, 1997.

Knell, Simon. *The Culture of English Geology, 1815–1851*. Aldershot: Ashgate Publishing, 2000.

Knowles, James. "Aspects of Tennyson." *Nineteenth Century* 33 (1893): 164–188.

Koerner, E. F. Konrad. "William Dwight Whitney and the influence of Geology on Linguistic Theory in the 19th Century." In *Language and Earth*, edited by Bernd Naumann, Franz Plank, and Gottfried Hofbauer. Amsterdam: Benjamins, 1992.

Kurata, Marilyn J. "'A Juggle Born of the Brain': A New Reading of *Maud*." *Victorian Poetry* 21 (1983): 366–378.

Lang, Cecil Y., and Edgar F. Shannon Jr., eds. *The Letters of Alfred Lord Tennyson II 1851–1870*, 3 vols. Oxford: Clarendon, 1987.

Laudan, Larry. *Science and Hypothesis: Historical Essay on Scientific Methodology*. Holland, Boston, and London: D. Reidel, 2010.

Levine, George. *Darwin and the Novelists: Patterns of Science in Victorian Fiction*. Chicago: University of Chicago Press, 1988.

————. ed. *One Culture: Essays in Science and Literature*. Wisconsin: University of Wisconsin Press, 1987.

Lodge, David. *After Bakhtin: Essays on Fiction and Criticism*. London: Routledge, 1990.

Lougy, Robert E. "The Sounds and Silence of Madness: Language as Theme in Tennyson's *Maud*." *Victorian Poetry* 22, no. 4 (1984): 407–426.

Lutz, Deborah. "The Dead Still Among Us: Victorian Secular Relic, Hair Jewellery, and Death Culture." *Victorian Literature and Culture* 39, no. 1 (2011): 127–142.

————. *Relics of Death in Victorian Literature and Culture*. Cambridge: Cambridge University Press, 2015.

Lyell, Charles. "Scientific Institutions." *Quarterly Review* 34 (June–Sept. 1826): 153–179.

————. "Transactions of the Geological Society." *Quarterly Review* 34 (June–Sept. 1826): 507–540.

————. "State of the Universities." *Quarterly Review* 36 (June–Oct. 1827): 216–268.

————. "Scrope's *Geology of Central France*." *Quarterly Review* 36 (June–Oct. 1827): 437–483. Review of George Poulett Scrope, "A Memoir on the Geology of Central France," 1827.

————. *Principles of Geology.* 1830–1833. 3 vols. Reprint of 1st edition, with an introduction by Martin Rudwick. Chicago: University of Chicago Press, 1990.

————. *Elements of Geology.* London: John Murray, 1838.

————. *Travels in North America,* 2 vols., London: Murray, 1845.

Lyell, Katherine, ed. *Life Letters and Journals of Sir Charles Lyell, Bart,* 2 vols. London: John Murray, 1881.

Macovski, Michael, ed. *Dialogue and Critical Discourse: Language, Culture, Critical Theory.* Oxford, Oxford University Press, 1997.

Mangles, James Henry. *Tennyson at Aldworth: The Diary of James Henry Mangles,* edited by Earl A. Knies. Athens: University of Ohio Press, 1984.

Mann, Robert James. *Tennyson's 'Maud' Vindicated: An Explanatory Essay.* London: Jarrold & Sons, 1856.

Mantell, Gideon. *The Wonders of Geology Or, A Familiar Exposition of Geological Phenomena,* 2 vols. London: Relfe and Fletcher, 1838.

Marshall, Nancy Rose. ""A Dim World, Where Monsters Dwell": The Spatial Time of the Sydenham Crystal Palace Dinosaur Park." *Victorian Studies* 49, no. 2 (2007): 286–301.

Marston, J. W. (unsigned review). "The Princess (1847)." *The Athenaeum* 21 (January 1, 1848): 6–8.

Martin, Robert Bernard. *Tennyson: The Unquiet Heart.* Oxford: Clarendon Press, 1980.

Mason, Michael. "The Timing of *In Memoriam.*" In *Studies in Tennyson,* edited by Hallam Tennyson. London and Basingstoke: McMillan, 1981.

Mattes, Eleanor. Bustin. *In Memoriam: The Way of the Soul.* New York: Exposition Press, 1951.

McDonald, Peter. *Sound Intentions: The Workings of Rhyme in Nineteenth-Century Poetry.* Oxford: Oxford University Press, 2012.

McPhee, John. *Basin and Range.* New York: Farrar, Straus & Giroux, 1981.

Miller, Hugh, *Hugh Miller's Memoir: From Stonemason to Geologist,* edited by Michael Shortland. Edinburgh: Edinburgh University Press, 1995.

————. *The Old Red Sandstone; or, New Walks in an Old Field.* 1841, 3rd edition. Edinburgh: John Johnson, 1847.

Millhauser, Milton. "Tennyson's *Princess* and *Vestiges.*" *Modern Language Association* 69, no. 1 (1954): 337–343.

————. *Just Before Darwin: Robert Chambers and Vestiges.* Connecticut: Wesleyan University Press, 1959.

————. "Tennyson, *Vestiges,* and the Dark Side of Science." *Victorian Newsletter* 17 (1969): 22–25.

————. *Fire and Ice: The Influence of Science on Tennyson's Poetry.* Lincoln: Tennyson Research Centre, 1971.

Murray, John IV. *John Murray III, 1808–1892: A Brief Memoir*. London: Murray, 1919.

Naumann, Bernd. "History of the Earth and Origin of Language in the 18th and 19th Century: The Case for Catastrophism." In *Language and Earth*, edited by Bernd Naumann et al. Amsterdam: John Benjamins, 1992.

Nicolson, Marjorie. *Mountain Gloom, Mountain Glory: The Development of the Aesthetics of the Infinite*. Seattle: University of Washington Press, 1959.

O'Connor, Ralph. *The Earth on Show: Fossils and the Poetics of Popular Science, 1802–56*. Chicago and London: University of Chicago Press, 2007.

———. "The meaning of 'literature' and the place of modern scientific nonfiction in Literature and Science." *Journal of Literature and Science* 10, no. 1 (2017): 37–45.

Ospovat, Alexander. "The Distortion of Werner in Lyell's *Principles of Geology*." *British Journal for the History of Science* 9 (1976): 190–198.

Owen, Richard. *Lectures on the Comparative Anatomy and Physiology of the Vertebrate Animals*, 2 vols. London: Longman, Brown, Green and Longmans, 1846.

———. *Geology and Inhabitants of the Ancient World*. London: Crystal Palace Library and Bradbury & Evans, 1854.

———. *The Life of Richard Owen*. London: John Murray, 1894.

Pease, Allison. "Maud and its Discontents." *Criticism: A Quarterly for Literature and the Arts* 36, no. 1 (1994): 101–118.

Piggott, Jan. *Palace of the People: The Crystal Palace at Sydenham 1854–1939*. London: Hurst, 2004.

Playfair, John. *Illustrations of the Huttonian Theory of the Earth*. 1802. New York: Dover Publications, 1956.

Porter, Roy. "Charles Lyell and the Principles of the History of Geology." *British Journal for the History of Science* 9 (1976): 91–103.

———. "Philosophy and Politics of a Geologist: G. H. Toulmin (1754–1817)." *Journal of the History of Ideas* 39, no. 3 (1978): 435–449.

Potter, George R. "Tennyson and the Biological Theory of Mutability in Species." *Philosophical Quarterly* 16 (1939): 321–343.

Priestley, F.E.L. *Language and Structure in Tennyson's Poetry*. London: Andre Deutsch, 1973.

Rader, Ralph. *Tennyson's "Maud": The Biographical Genesis*. Berkeley: University of California Press, 1963.

Raub, Cymbre Quincy. "Robert Chambers and William Whewell: A Nineteenth-Century Debate Over the Origin of Language." *Journal of the History of Science* 49, no. 3 (April–June. 1988): 287–300.

Ricks, Christopher. *Tennyson*. Basingstoke: Macmillan, 1989.

———. ed. *Tennyson: A Selected Edition Incorporating the Trinity College Manuscripts*. Berkeley and Los Angeles: University of California Press, 1989.

Rossi, Paolo. *The Dark Abyss of Time: The History of the Earth & the History of Nations From Hooke to Vico*. Chicago: University of Chicago Press, 1984.

Rudwick, Martin. "The Strategy of Lyell's *Principles of Geology*." *Isis* 61 (1970): 5–33.

———. *The Meaning of Fossils, Episodes in the History of Palaeontology*. Chicago: University of Chicago Press, 1972.

———. *Georges Cuvier, Fossil Bones, and Geological Catastrophes*. Chicago: University of Chicago Press, 1997.

———. Martin *Bursting the Limits of Time: The Reconstruction of Geohistory in an Age of Revolution*. Chicago: Chicago University Press, 2005.

———. *Worlds Before Adam: The Reconstruction of Geohistory in the Age of Reform*. Chicago, London: University of Chicago Press, 2008.

Rupke, Nicolaas. *The Great Chain of Being*. Oxford: Oxford University Press, 1983.

———. *Richard Owen: Biology Without Darwin, a Revised Edition*. 2009. Chicago: University of Chicago Press, 1994.

Ruse, Michael. *The Darwinian Revolution: Science Red in Tooth and Claw*. Chicago: University of Chicago Press, 1999.

Ruskin, John. *The Works of John Ruskin*, 39 vols. London: George Allen, 1912.

Rutland, W. R. "Tennyson and the Theory of Evolution." *Essays and Studies* 26 (1940): 7–29.

Schiebinger, Londa. "Why Mammals are Called Mammals: Gender Politics in Eighteenth-Century Natural History." *American Historical Review* 98, no. 2 (1998): 382–411.

Secord, James, ed. *Charles Lyell, Principles of Geology*. 1830–33. London: Penguin, 1997.

———. *Victorian Sensation: The Extraordinary Publication, Reception, and Secret Authorship of Vestiges of the Natural History of Creation*. Chicago and London: University of Chicago Press, 2003.

———. *Visions of Science: Books and Readers at the Dawn of the Victorian Age*. Oxford: Oxford University Press, 2014.

Sedgwick, Adam. "Vestiges of the Natural History of Creation." *Edinburgh Review*, CLXV (1845).

Sedgwick, Eve Kosofsky. "Tennyson's Princess: One Bride for Seven Brothers." (1985) In *Tennyson*, edited by Rebecca Stott. London: Longman, 1996.

Shannon, Edgar Finley., *Tennyson and the Reviewers: A Study of His Reputation and of the Influences of the Critics Upon His Poetry 1827–1851*. Cambridge, MA: Harvard University Press, 1952.

Shatto, Susan, and Marion Shaw, eds. *Tennyson: In Memoriam*. Oxford: Clarendon Press, 1982.

Shatto, Susan, ed. *Tennyson's Maud: A Definitive Edition*. London: Althone Press, 1986.

Shaw, David. *Tennyson's Style*. Ithaca and London: Cornell University Press, 1976.

Shaw, Marion. *Alfred Lord Tennyson*. London: Harvester Wheatsheaf, 1988.

Shires, Linda M., "Maud, Masculinity and Poetic Identity." *Criticism: A Quarterly for Literature and the Arts* 29, no. 3 (1987): 269–290.

Sinfield, Alan. *The Language of Tennyson's In Memoriam*. New York: Barnes & Noble, 1971.

———. *Alfred Tennyson*. Oxford: Blackwell, 1986.

Sleigh, Charlotte. *Literature and Science*. Basingstoke: Palgrave Macmillan, 2011.

Slinn, E. Warwick. *The Discourse of Self in Victorian Poetry*. London: Macmillan, 1991.

———. *Victorian Poetry as Cultural Critique*. Virginia: University of Virginia Press, 2003.

Smiles, Samuel. *Self-Help*. London: Murray, 1859.

Stitt, Megan Perigoe. *Metaphors of Change in the Language of Nineteenth-century Fiction: Scott, Gaskell, and Kingsley*. Oxford: Clarendon Press, 1998.

Stott, Rebecca, ed. *Tennyson*. London: Longman, 1996.

———. "Tennyson's Drift: Evolution in *The Princess*." In *Darwin, Tennyson and Their Readers*, edited by Valerie Purton. London and New York: Anthem Press, 2014.

Tate, Gregory. *The Poet's Mind: The Psychology of Victorian Poetry*. Oxford: Oxford University Press, 2012.

———. "The Uses of Poetry in Science," November 2015: https://drgregorytate.wordpress.com/2015/11/30/the-uses-of-poetry-in-victorian-science/.

Taylor, Barbara. *Eve and the New Jerusalem*. London: Virago, 1983.

Taylor, M. A., and L. I. Anderson. "Tennyson and the Geologists Part 2: Saurians and the Isle of Wight." *Tennyson Research Bulletin* 10, no. 5 (2016), in press.

Tennyson, Alfred. *The Poems of Tennyson*, edited by Christopher Ricks. London: Longman, 1969.

———. *Letters of Alfred Lord Tennyson, I, 1821–1850*, edited by Lang Cecil Y. and Edgar F. Shannon, Jr. Oxford: Clarendon Press, 1982.

Tennyson, Charles. *Alfred Tennyson*. London: Macmillan, 1949.

Tennyson, Hallam, T. *Alfred Lord Tennyson: A Memoir by his Son*, 2 vols, London: Macmillan, 1897.

———, ed. *Tennyson and his Friends*. London: Macmillan, 1911.

Thesing, William B. "Tennyson and the City: Historical Tremours and Hysterical Tremblings." *Tennyson Research Bulletin* 3, no. 1 (1977): 14–22.

Timko, Michael. *Tennyson and Carlyle*. Iowa: University of Iowa, 1987.

Todorov, Tzvetan. *Mikhail Bakhtin: The Dialogic Principle*. Translated by Wlad Godzich. Manchester: Manchester University Press, 1984.

Tomaiuolo, Saverio. "Tennyson and the Crisis of Narrative Voice in *Maud*." *Tennyson Research Bulletin* 8, no. 1 (2002): 28–43.

Tomko, Michael. "Varieties of Geological Experience: Religion, Body, and Spirit in Tennyson's *In Memoriam* and Lyell's *Principles of Geology*." *Victorian Poetry* 2, no. 2 (2004): 113–133.

Toulmin, George Hoggart. *The Eternity of the Universe*. London: J. Johnson, 1789.

Trench, R. C. *On the Study of Words: Five Lectures Addressed to the Pupils at the Diocesan Training School, Winchester* (1851). London: John W. Parker and Son, 1853.

———. *English: Past and Present, Five Lectures*. London: John W. Parker and Son, 1855.

———. *A Select Glossary of English Words Used Formerly in Senses Different From Their Present*. London: John W. Parker and Son, 1859.

Tucker, Herbert. *Tennyson and the Doom of Romanticism*. Cambridge, MA: Harvard University Press, 1988.

Turner, Paul. *Tennyson: Routledge Author Guides*. London: Routledge & Kegan Paul, 1976.

Ulrich, John M. "Thomas Carlyle, Richard Owen, and the Paleontological Articulation of the Past." *Journal of Victorian Culture* 11, no. 1 (2006): 30–58.

Van Dyke, Henry. *Studies in Tennyson*. New York: Charles Scribner, 1920.

Vincent, David. *Bread, Knowledge and Freedom: A Study of Nineteenth-Century Working Class Autobiography*. New York: Methuen, 1981.

Warren, Thomas Herbert. "FitzGerald, Carlyle, and Other Friends." In *Tennyson and his Friends*, edited by Hallam T. Tennyson. London: Macmillan, 1911.

Wheeler, Michael. *Heaven, Hell and the Victorians*. Cambridge: Cambridge University Press, 1990.

Whewell, William. *Indications of the Creator*. 1845. 2nd edition. London: John W. Parker, 1846.

———. *History of the Inductive Sciences*. 1837. 3 vols. Facsimile reprint of 1857 edition. London: Frank Cass, 1967.

———. "On the Fundamental Antithesis of Philosophy." *Transactions of the Cambridge Philosophical Society* 8 (1949).

———. *Of the Plurality of Worlds*. London: John W. Parker, 1853.

Whitney, William Dwight, *Language and the Study of Language: Twelve Lectures on the Principle of Linguistic Science*. 1867. London: N. Trübner & Co., 1870.

Willey, Basil. "Tennyson." In *More Nineteenth Century Studies*. London: Chatto & Windus, 1956.

Willis, Martin. *Literature and Science*. London: Palgrave Macmillan, 2015.

Wilson, Edward, O. *Consilience: the Unity of Knowledge.* London: Little, Brown and Company, 1998.

Wilson, Leonard. *Charles Lyell the Years to 1841: The Revolution in Geology.* New Haven and London: Yale University Press, 1972.

Wood, Robert Muir. *The Dark Side of the Earth.* London: Allen & Unwin, 1985.

Wordsworth, Jonathan. "'What is it, that has been done?': The Central Problem of *Maud.*" *Essays in Criticism* 24 (1974): 356–362.

Wyatt, John. *Wordsworth and the Geologists.* Cambridge: Cambridge University Press, 1995.

Zimmerman, Virginia. *Excavating Victorians.* New York: New York State University Press, 2008.

INDEX

© The Editor(s) (if applicable) and The Author(s) 2017   215
M. Geric, *Tennyson and Geology*, Palgrave Studies in Literature,
Science and Medicine, DOI 10.1007/978-3-319-66110-0